KUMON MATH WORKBOOKS

Geometry & Measurement

Table of Contents

KUMON

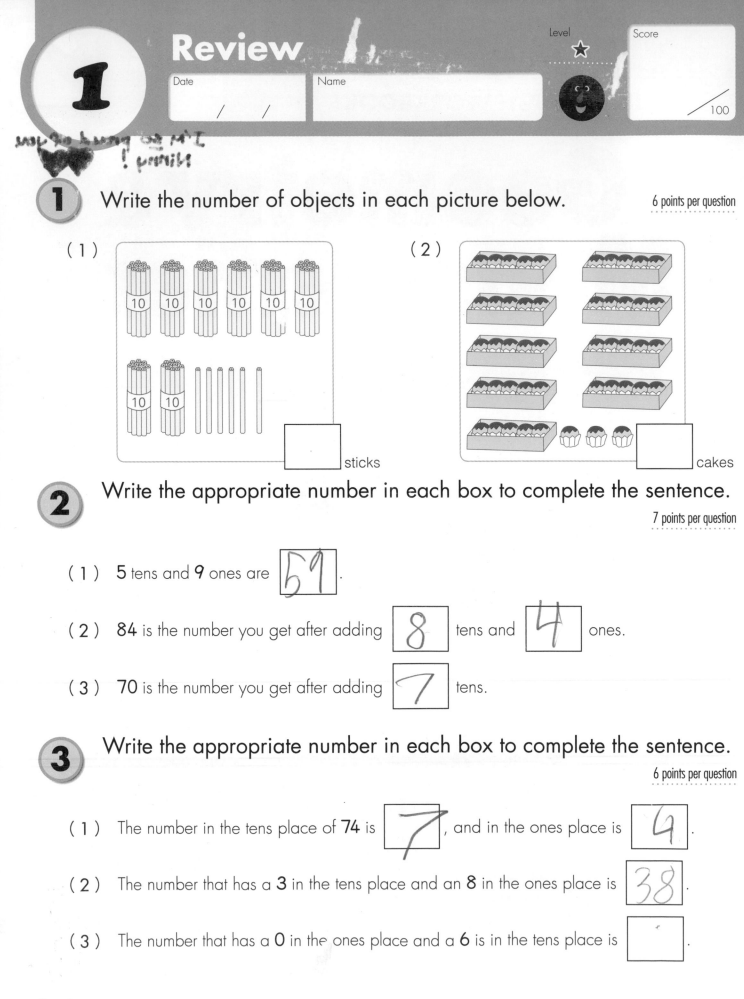

1 Review

Date   /   /

Name

Level ☆

Score   /100

1 Write the number of objects in each picture below.

6 points per question

(1)

| 10 | 10 | 10 | 10 | 10 | 10 |

| 10 | 10 |

sticks

(2)

cakes

2 Write the appropriate number in each box to complete the sentence.

7 points per question

(1) 5 tens and 9 ones are 59 .

(2) 84 is the number you get after adding 8 tens and 4 ones.

(3) 70 is the number you get after adding 7 tens.

3 Write the appropriate number in each box to complete the sentence.

6 points per question

(1) The number in the tens place of 74 is 7 , and in the ones place is 4 .

(2) The number that has a 3 in the tens place and an 8 in the ones place is 38 .

(3) The number that has a 0 in the ones place and a 6 is in the tens place is ⬚ .

2   © Kumon Publishing Co., Ltd.

4 Draw a line from 70 to 100 in order.

15 points for completion

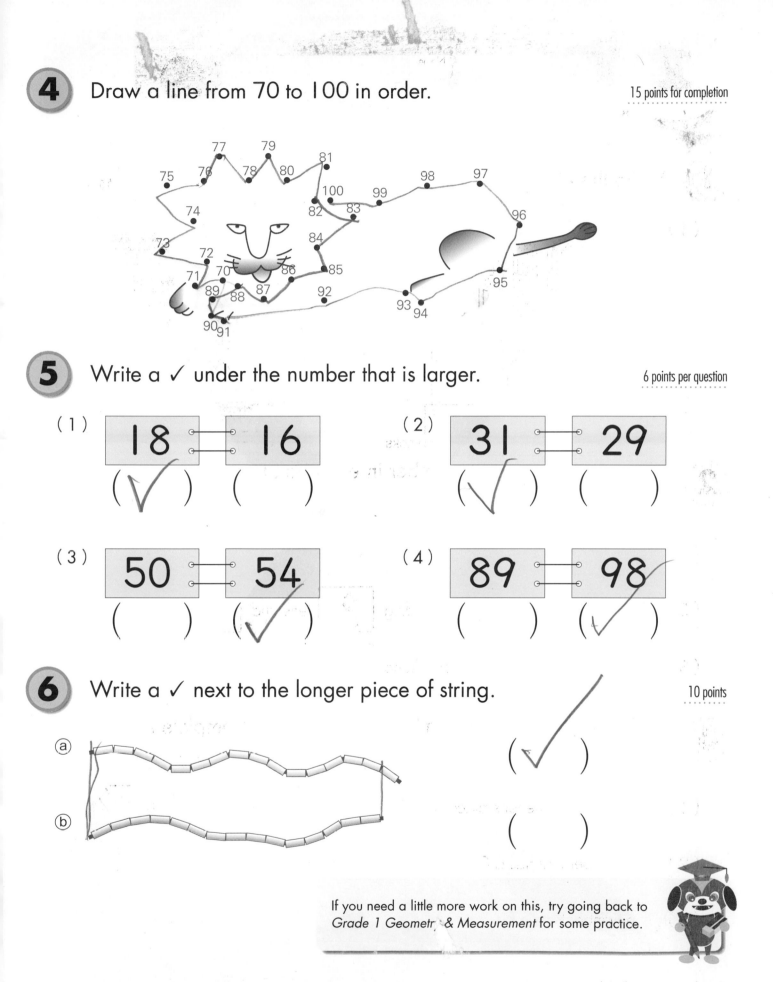

5 Write a ✓ under the number that is larger.

6 points per question

(1) | 18 | 16 |
(✓) ( )

(2) | 31 | 29 |
(✓) ( )

(3) | 50 | 54 |
( ) (✓)

(4) | 89 | 98 |
( ) (✓)

6 Write a ✓ next to the longer piece of string.

10 points

ⓐ

(✓)

ⓑ

( )

If you need a little more work on this, try going back to *Grade 1 Geometry & Measurement* for some practice.

1 Fill in the missing number in each box on the number lines below.

3 points per box

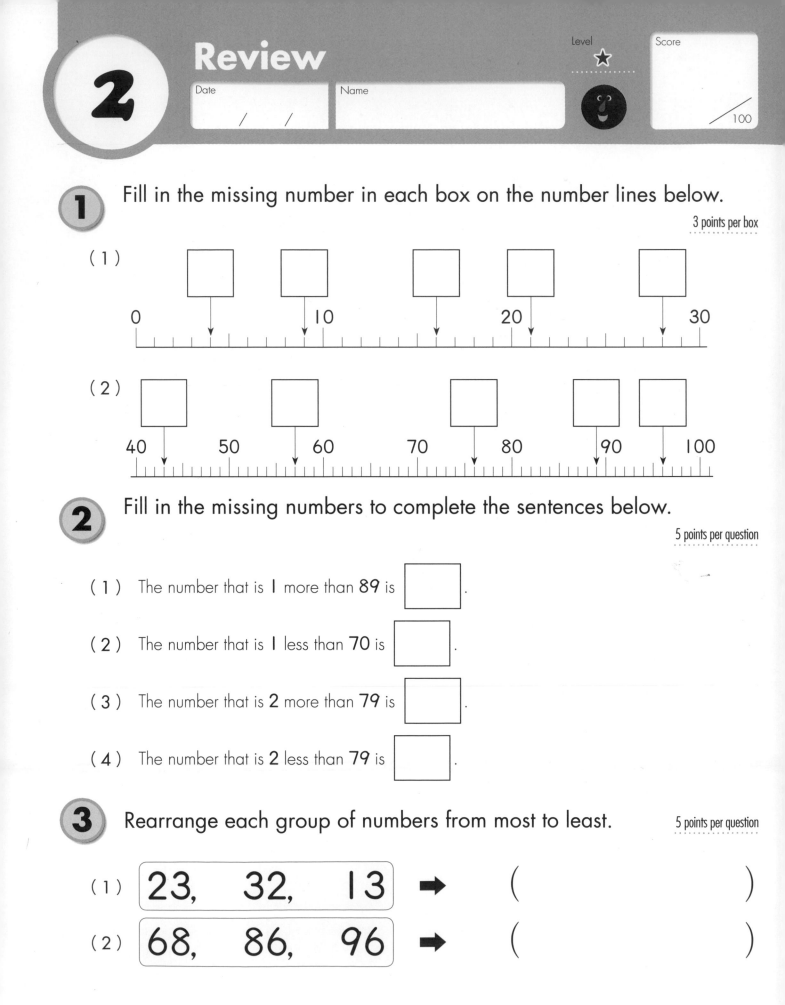

(1)

0   □   10   □   20   □   30

(2)

□   40   □   50   □   60   70   □   80   □   90   □   100

2 Fill in the missing numbers to complete the sentences below.

5 points per question

(1) The number that is 1 more than 89 is □.

(2) The number that is 1 less than 70 is □.

(3) The number that is 2 more than 79 is □.

(4) The number that is 2 less than 79 is □.

3 Rearrange each group of numbers from most to least.

5 points per question

(1) 23, 32, 13 ➡ (   )

(2) 68, 86, 96 ➡ (   )

4 Fill in the missing number in each box on the number lines below.

5 points per question

(1) ☐ — 37 — 38 — ☐ — 40 — ☐ — 42 — 43

(2) 53 — 52 — 51 — ☐ — ☐ — 48 — ☐ — 46

(3) 30 — ☐ — 50 — 60 — ☐ — 80 — 90 — ☐

(4) 22 — 32 — ☐ — 52 — 62 — ☐ — 82 — ☐

5 Which pencil is longer? How many units longer is it?

10 points

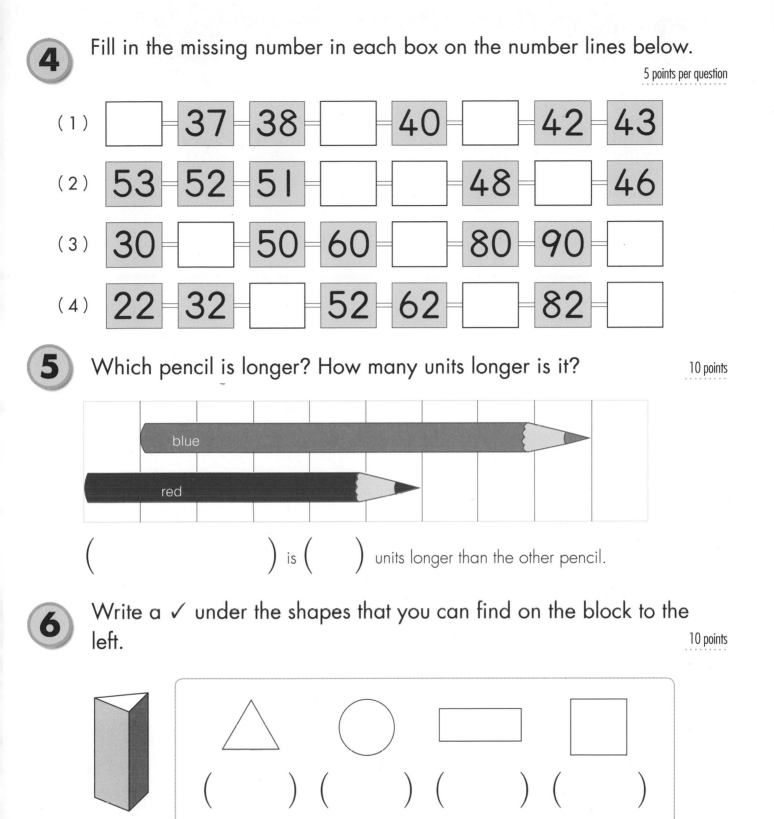

blue

red

(_____) is (_____) units longer than the other pencil.

6 Write a ✓ under the shapes that you can find on the block to the left.

10 points

(_____) (_____) (_____) (_____)

Remember to check the pattern before completing the number lines!

3 Numbers up to 1,000 ★★

Date   /   /

Name

Score   /100

1 Write the number of sticks in each box below.   5 points per question

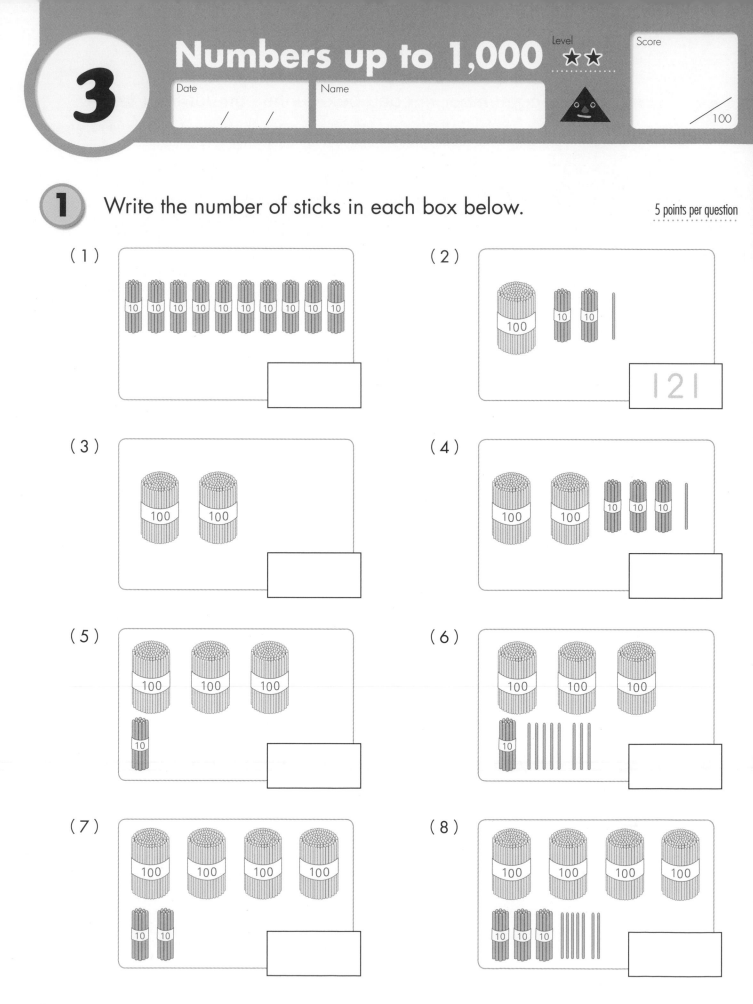

(1)

(2)   121

(3)

(4)

(5)

(6)

(7)

(8)

© Kumon Publishing Co., Ltd.

2 How many objects are in each picture? Write the number in each box below.

15 points per question

(1)

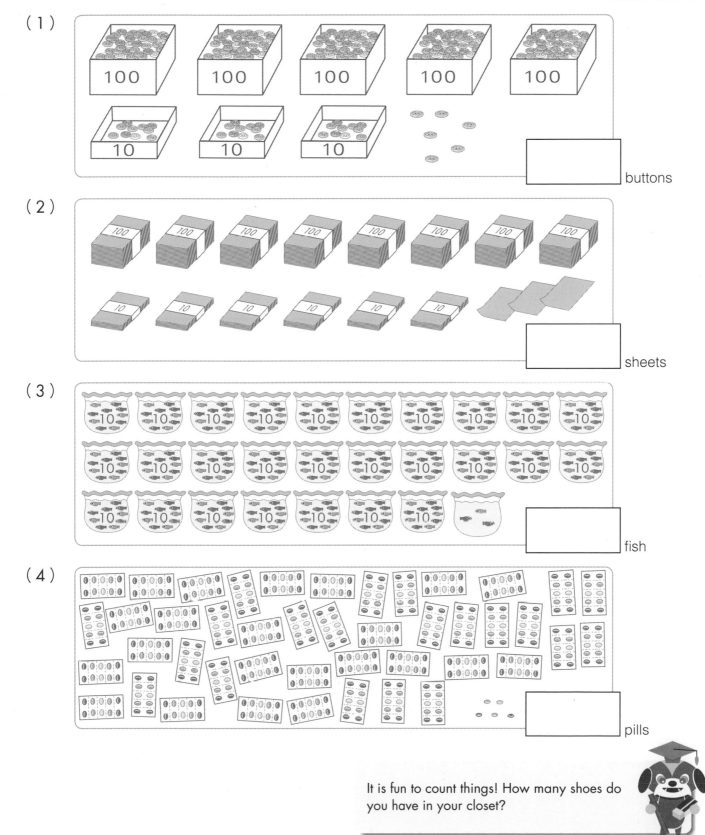

buttons

(2)

sheets

(3)

fish

(4)

pills

It is fun to count things! How many shoes do you have in your closet?

1 Fill in the missing numbers to complete the sentences below.

5 points per question

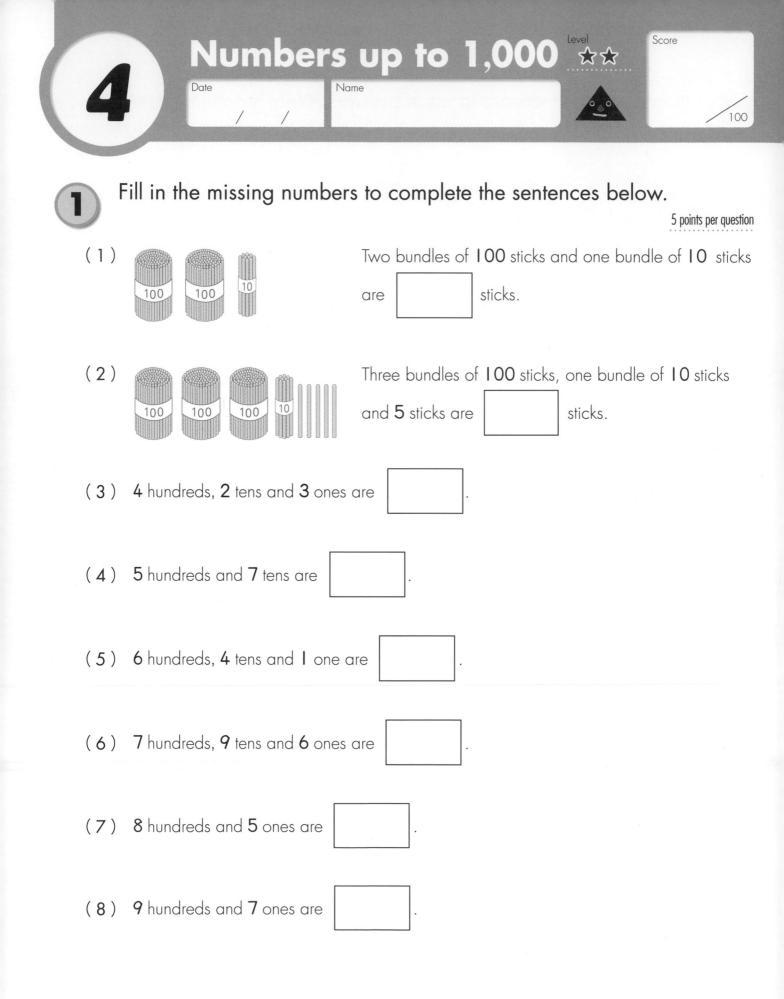

(1) Two bundles of 100 sticks and one bundle of 10 sticks are ☐ sticks.

(2) Three bundles of 100 sticks, one bundle of 10 sticks and 5 sticks are ☐ sticks.

(3) 4 hundreds, 2 tens and 3 ones are ☐ .

(4) 5 hundreds and 7 tens are ☐ .

(5) 6 hundreds, 4 tens and 1 one are ☐ .

(6) 7 hundreds, 9 tens and 6 ones are ☐ .

(7) 8 hundreds and 5 ones are ☐ .

(8) 9 hundreds and 7 ones are ☐ .

2 Fill in the missing numbers or words to complete the sentences below.

6 points per question

(1) 250 is the number you get after adding ☐ hundreds and ☐ tens.

(2) 485 is the number you get after adding ☐ hundreds, ☐ tens, and ☐ ones.

(3) 807 is the number you get after adding ☐ hundreds and 7 ☐ .

(4) 5 hundreds are ☐ .

(5) 10 hundreds are ☐ .

(6) 1,000 is the number you get after adding ☐ hundreds.

(7) 760 is the number you get after adding ☐ tens.

3 Write the appropriate number in each box below.

9 points per question

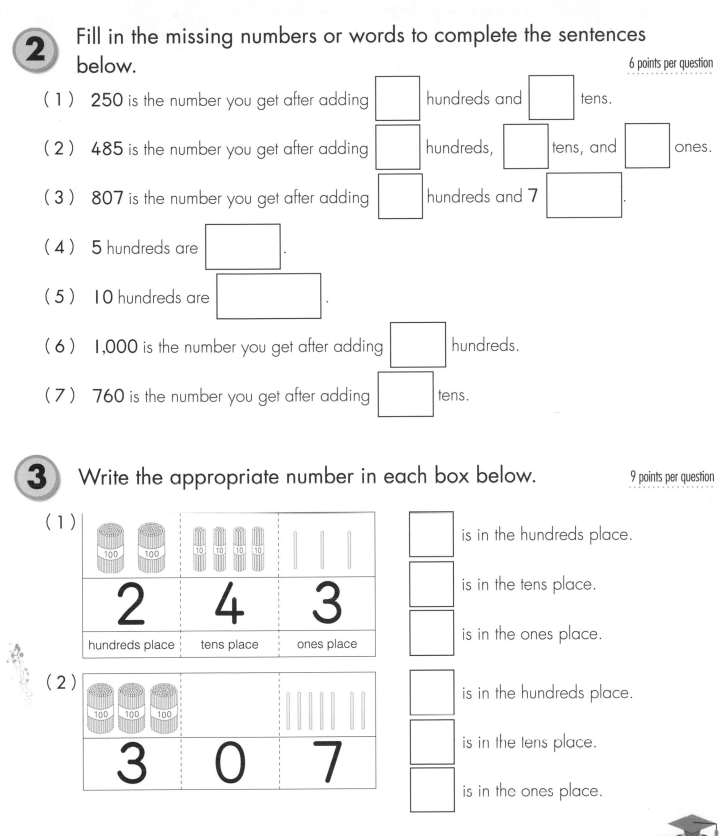

(1)

2	4	3
hundreds place	tens place	ones place

☐ is in the hundreds place.

☐ is in the tens place.

☐ is in the ones place.

(2)

3	0	7

☐ is in the hundreds place.

☐ is in the tens place.

☐ is in the ones place.

Looks like you are getting good at using big numbers. Good job!

Numbers up to 1,000

Date / /

Name

Score /100

1 Fill in the missing number in each box on the number lines below.

2 points per box

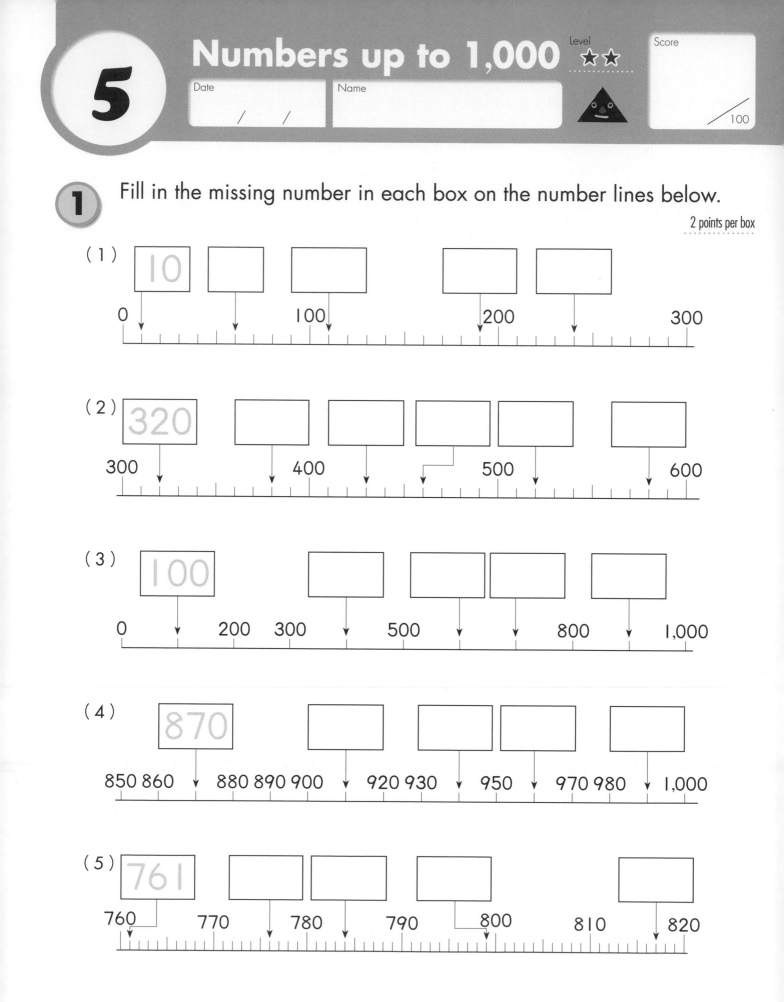

(1)

| 10 | | | | |

0   100   200   300

(2)

| 320 | | | | | |

300   400   500   600

(3)

| 100 | | | | |

0   200 300   500   800   1,000

(4)

| 870 | | | | |

850 860   880 890 900   920 930   950   970 980   1,000

(5)

| 761 | | | | |

760   770   780   790   800   810   820

2 Using the number line as a guide, fill in each box to complete the sentences below.

6 points per question

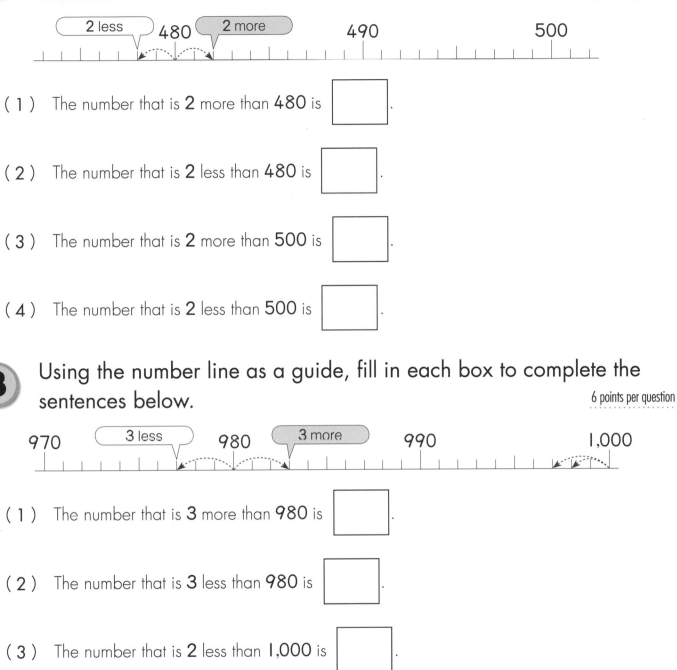

(1) The number that is **2** more than **480** is ☐.

(2) The number that is **2** less than **480** is ☐.

(3) The number that is **2** more than **500** is ☐.

(4) The number that is **2** less than **500** is ☐.

3 Using the number line as a guide, fill in each box to complete the sentences below.

6 points per question

(1) The number that is **3** more than **980** is ☐.

(2) The number that is **3** less than **980** is ☐.

(3) The number that is **2** less than **1,000** is ☐.

(4) The number that is **3** less than **1,000** is ☐.

Now the numbers are going to get even bigger!

1 Using the number line as a guide, fill in each box to complete the sentences below.

5 points per question

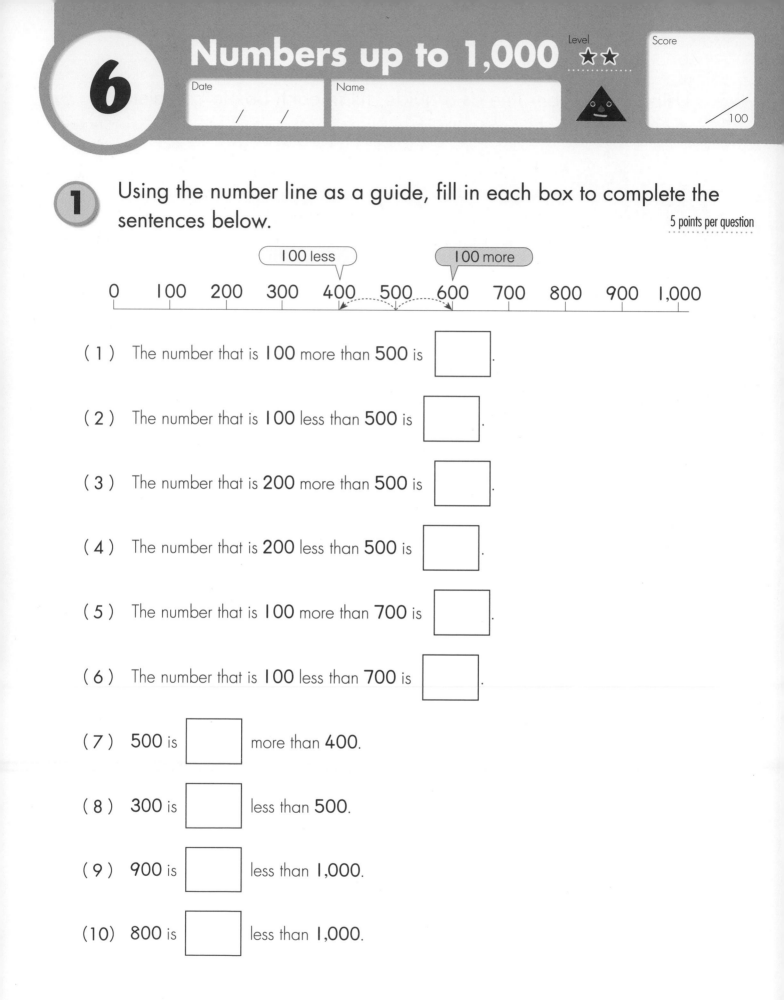

(1) The number that is 100 more than 500 is ☐ .

(2) The number that is 100 less than 500 is ☐ .

(3) The number that is 200 more than 500 is ☐ .

(4) The number that is 200 less than 500 is ☐ .

(5) The number that is 100 more than 700 is ☐ .

(6) The number that is 100 less than 700 is ☐ .

(7) 500 is ☐ more than 400.

(8) 300 is ☐ less than 500.

(9) 900 is ☐ less than 1,000.

(10) 800 is ☐ less than 1,000.

2 Using the number line as a guide, fill in each box to complete the sentences below.

5 points per question

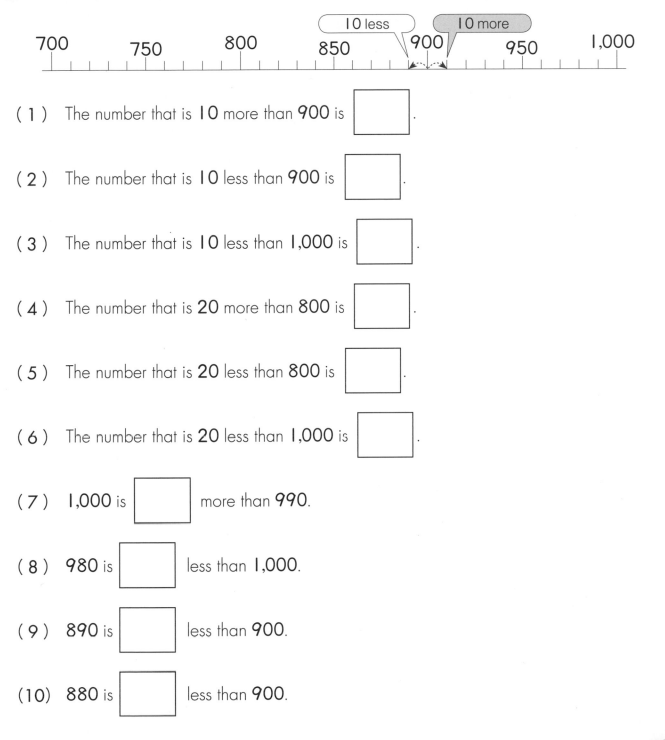

700   750   800   850 | 10 less | 900 | 10 more | 950   1,000

(1) The number that is 10 more than 900 is ☐ .

(2) The number that is 10 less than 900 is ☐ .

(3) The number that is 10 less than 1,000 is ☐ .

(4) The number that is 20 more than 800 is ☐ .

(5) The number that is 20 less than 800 is ☐ .

(6) The number that is 20 less than 1,000 is ☐ .

(7) 1,000 is ☐ more than 990.

(8) 980 is ☐ less than 1,000.

(9) 890 is ☐ less than 900.

(10) 880 is ☐ less than 900.

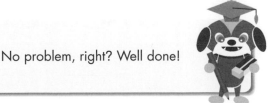

No problem, right? Well done!

7 Numbers up to 1,000

Level ★★

Date   /   /

Name

Score   /100

1 Write a ✓ under the larger number.

5 points per question

(1) 200 — 300
( )   ( )

(2) 700 — 500
( )   ( )

(3) 450 — 350
( )   ( )

(4) 180 — 280
( )   ( )

(5) 324 — 414
( )   ( )

(6) 298 — 306
( )   ( )

(7) 280 — 270
( )   ( )

(8) 365 — 395
( )   ( )

(9) 181 — 179
( )   ( )

(10) 467 — 476
( )   ( )

(11) 394 — 395
( )   ( )

(12) 508 — 502
( )   ( )

14   © Kumon Publishing Co., Ltd.

2 Answer the questions below.

(1) Circle the numbers that are more than 546.

246   846   146   646   446   746

(2) Circle the numbers that are more than 546.

596   556   506   536   586   526

(3) Circle the numbers that are more than 546.

544   547   549   545   540   546

(4) Circle the numbers that are more than 546.

946   516   660   548   836   346

(5) Circle the numbers that are less than 546.

564   456   551   545   499   654

(6) Circle the numbers that are less than 546.

526   576   486   636   398   539

(7) Circle the numbers that are more than 735.

835   705   798   627   585   935

(8) Circle the numbers that are less than 735.

753   573   726   743   685   805

(9) Circle the numbers that are more than 381.

391   371   400   298   318   521

(10) Circle the numbers that are less than 381.

318   831   379   388   400   299

Ready for another step up? Good.

Level ★★

Date / /

Name

Score /100

1 How many sticks are there? Write the number in each box. 20 points per question

(1)

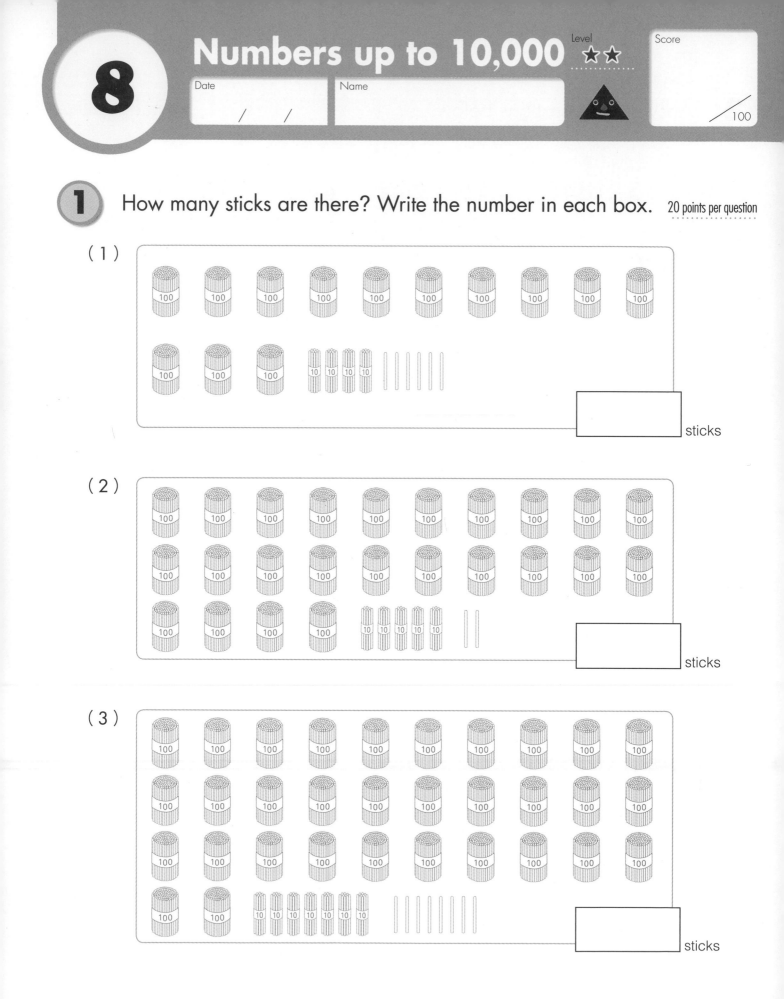

sticks

(2)

sticks

(3)

sticks

How many objects are there below? Write the number in each box.

10 points per question

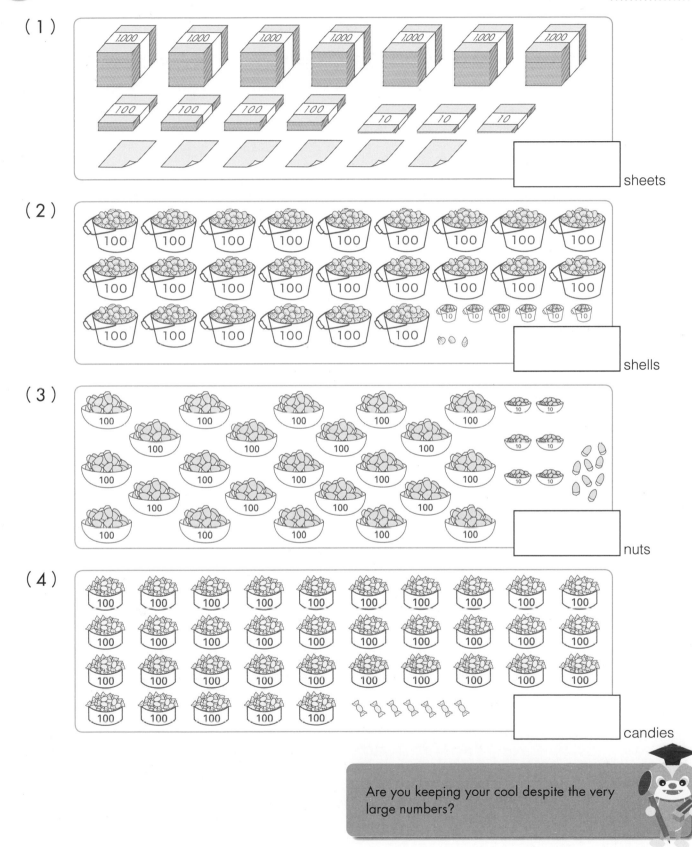

(1) sheets

(2) shells

(3) nuts

(4) candies

Are you keeping your cool despite the very large numbers?

17

Numbers up to 10,000

Date   /   /

Name

Score

/100

1 Fill in the missing numbers to complete the sentences below.

6 points per question

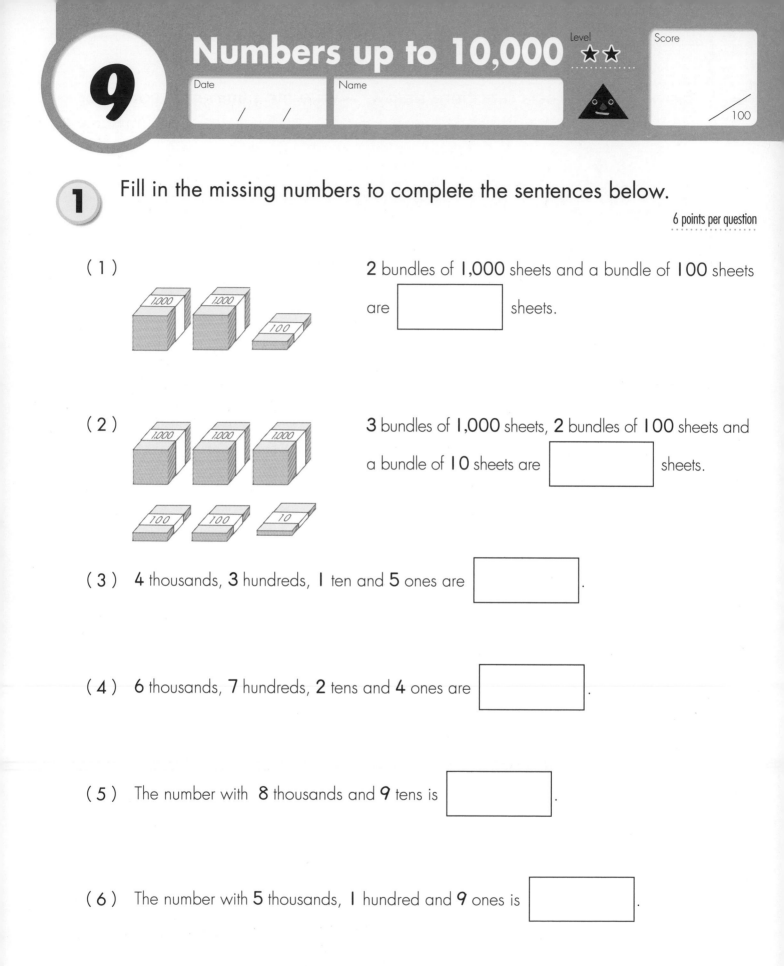

(1)

2 bundles of 1,000 sheets and a bundle of 100 sheets

are [   ] sheets.

(2)

3 bundles of 1,000 sheets, 2 bundles of 100 sheets and

a bundle of 10 sheets are [   ] sheets.

(3)   4 thousands, 3 hundreds, 1 ten and 5 ones are [   ] .

(4)   6 thousands, 7 hundreds, 2 tens and 4 ones are [   ] .

(5)   The number with 8 thousands and 9 tens is [   ] .

(6)   The number with 5 thousands, 1 hundred and 9 ones is [   ] .

2 Fill in the missing numbers or words to complete the sentences below.

6 points per question

(1) 3,400 is the result of adding [ ] thousands and [ ] hundreds.

(2) The result of adding 10 thousands is [ ].

(3) 1,500 is the result of adding [ ] hundreds.

(4) 10,000 is the result of adding [ ] thousands.

(5) 9,400 is the result of adding 9 [ ] and 4 [ ].

(6) 9,400 is the result of adding 94 [ ].

(7) 9,400 is the result of adding [ ] hundreds.

3 Write the appropriate number in each box below.

11 points per question

(1)

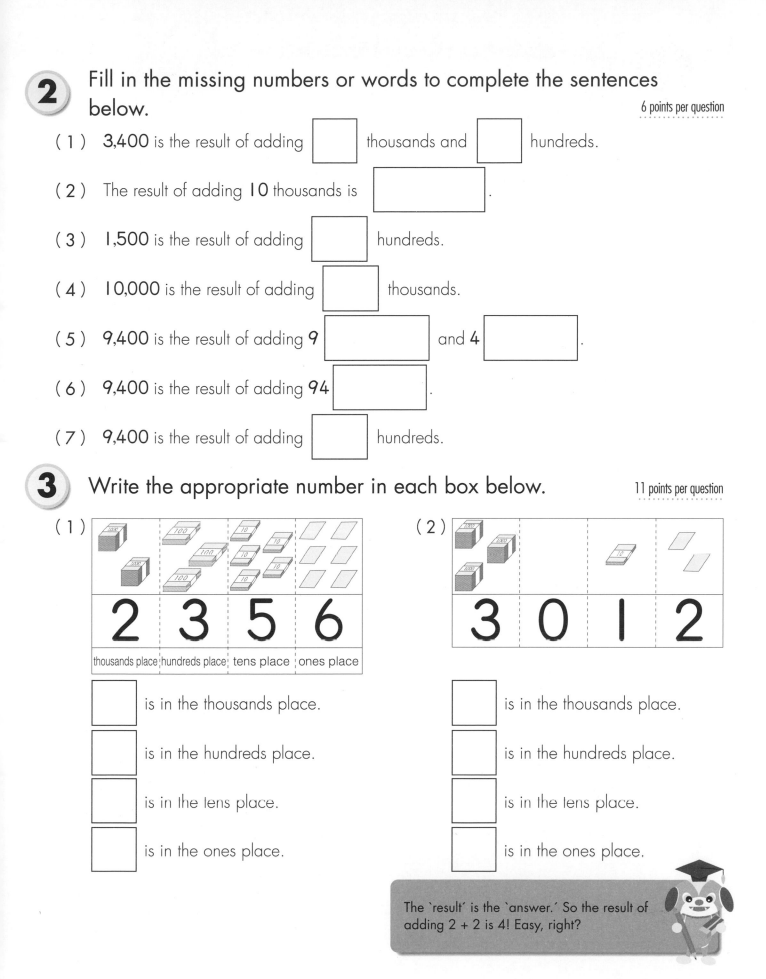

2	3	5	6
thousands place	hundreds place	tens place	ones place

[ ] is in the thousands place.

[ ] is in the hundreds place.

[ ] is in the tens place.

[ ] is in the ones place.

(2)

3	0	1	2

[ ] is in the thousands place.

[ ] is in the hundreds place.

[ ] is in the tens place.

[ ] is in the ones place.

The `result´ is the `answer.´ So the result of adding 2 + 2 is 4! Easy, right?

1 Fill in the missing number in each box on the number lines below.

2 points per box

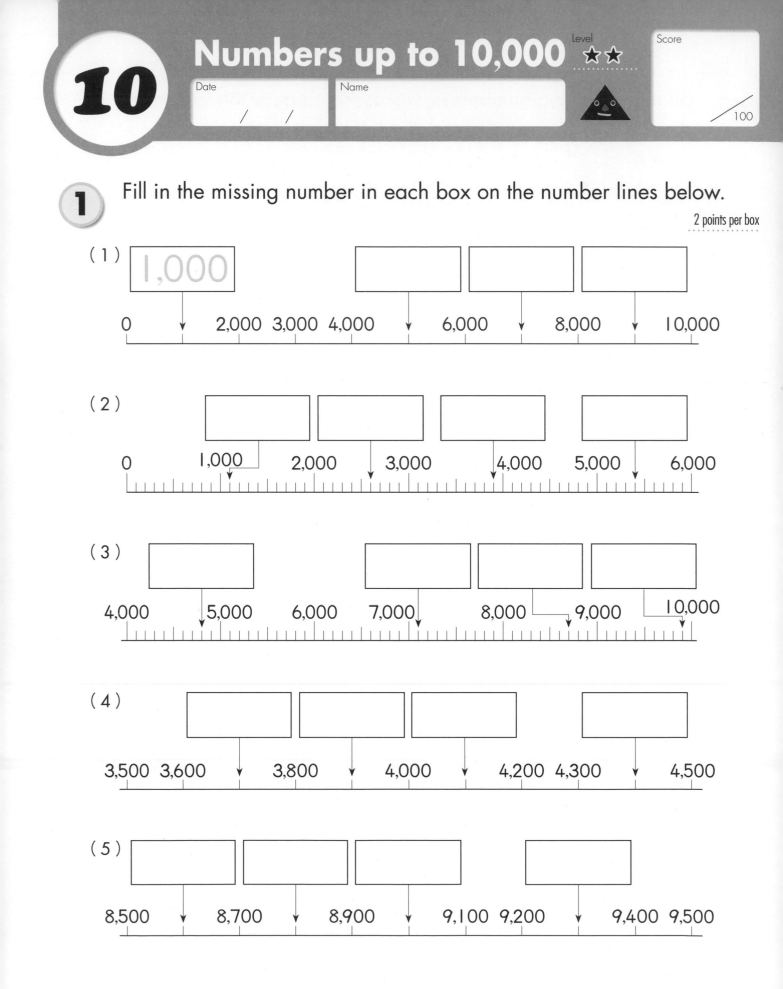

(1)
| 1,000 | | |

0   2,000 3,000 4,000   6,000   8,000   10,000

(2)

0   1,000   2,000   3,000   4,000   5,000   6,000

(3)

4,000   5,000   6,000   7,000   8,000   9,000   10,000

(4)

3,500 3,600   3,800   4,000   4,200 4,300   4,500

(5)

8,500   8,700   8,900   9,100 9,200   9,400 9,500

2 Fill in the missing number in each box on the number lines below.

2 points per box

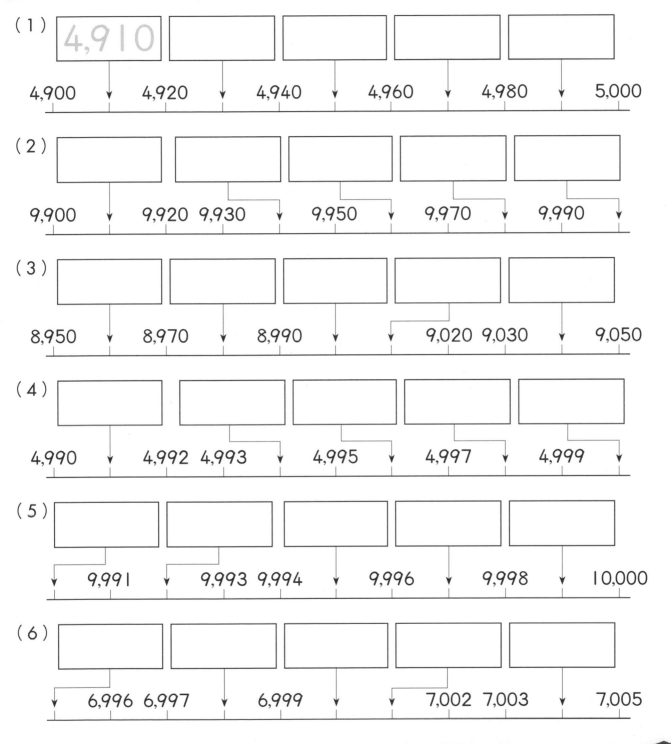

(1)
| 4,910 | | | | |

4,900   4,920   4,940   4,960   4,980   5,000

(2)
| | | | | |

9,900   9,920   9,930   9,950   9,970   9,990

(3)
| | | | | |

8,950   8,970   8,990   9,020   9,030   9,050

(4)
| | | | | |

4,990   4,992   4,993   4,995   4,997   4,999

(5)
| | | | | |

9,991   9,993   9,994   9,996   9,998   10,000

(6)
| | | | | |

6,996   6,997   6,999   7,002   7,003   7,005

Remember, practice makes perfect!

Numbers up to 10,000

Date   /   /

Name

Score
/ 100

1 Using the number line as a guide, fill in each box to complete the sentences below.

4 points per question

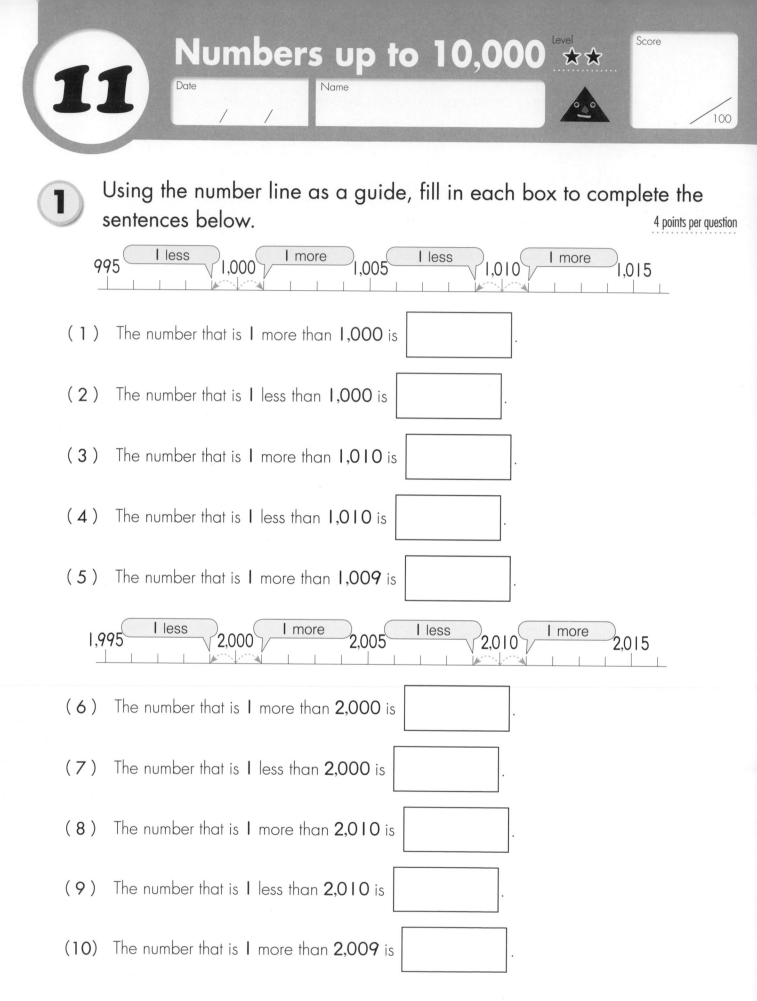

(1)   The number that is 1 more than 1,000 is [   ] .

(2)   The number that is 1 less than 1,000 is [   ] .

(3)   The number that is 1 more than 1,010 is [   ] .

(4)   The number that is 1 less than 1,010 is [   ] .

(5)   The number that is 1 more than 1,009 is [   ] .

(6)   The number that is 1 more than 2,000 is [   ] .

(7)   The number that is 1 less than 2,000 is [   ] .

(8)   The number that is 1 more than 2,010 is [   ] .

(9)   The number that is 1 less than 2,010 is [   ] .

(10)   The number that is 1 more than 2,009 is [   ] .

2 Fill in the missing numbers to complete the sentences below.

5 points per question

(1) The number that is 1 more than **5,000** is ☐.

(2) The number that is 1 less than **5,000** is ☐.

(3) The number that is 1 more than **5,010** is ☐.

(4) The number that is 1 more than **5,019** is ☐.

(5) The number that is 1 less than **7,000** is ☐.

(6) The number that is 1 less than **10,000** is ☐.

(7) The number that is 1 less than **9,010** is ☐.

(8) The number that is 1 more than **7,999** is ☐.

(9) The number that is 1 less than **6,000** is ☐.

(10) The number that is 1 less than **6,010** is ☐.

3 Using the number line as a guide, fill in each box to complete the sentences below.

5 points per question

0   1,000   2,000   3,000   4,000   5,000   6,000   7,000   8,000   9,000   10,000

(1) The number that is **1,000** more than **9,000** is ☐.

(2) The number that is **1,000** less than **9,000** is ☐.

You are getting good with big numbers!

Numbers up to 10,000 ★★

Level ★★

Date   /   /

Name

Score   /100

1   Write a ✓ under the larger number.

5 points per question

(1) 2,000 — 3,000
(   )   (   )

(2) 5,000 — 4,000
(   )   (   )

(3) 3,500 — 4,500
(   )   (   )

(4) 6,900 — 9,600
(   )   (   )

(5) 5,400 — 5,200
(   )   (   )

(6) 8,430 — 8,530
(   )   (   )

(7) 2,560 — 2,600
(   )   (   )

(8) 3,450 — 3,460
(   )   (   )

(9) 8,090 — 8,030
(   )   (   )

(10) 6,775 — 6,765
(   )   (   )

(11) 4,595 — 4,594
(   )   (   )

(12) 7,206 — 7,209
(   )   (   )

2 Answer the questions below.

(1) Circle the numbers that are more than 5,465.

7,465   2,465   6,465   5,465   8,465   3,465

(2) Circle the numbers that are more than 5,465.

5,165   5,565   5,465   5,965   5,865   5,365

(3) Circle the numbers that are more than 5,465.

5,435   5,475   5,425   5,485   5,465   5,445

(4) Circle the numbers that are more than 5,465.

5,469   5,461   5,468   5,467   5,464   5,466

(5) Circle the numbers that are less than 5,465.

5,765   4,565   5,290   5,457   3,995   6,545

(6) Circle the numbers that are more than 7,350.

8,350   7,035   7,620   6,985   5,490   7,360

(7) Circle the numbers that are less than 7,350.

5,730   7,150   7,290   6,850   7,400   8,150

(8) Circle the numbers that are more than 6,984.

6,990   6,094   6,890   8,496   5,992   7,014

(9) Circle the numbers that are less than 6,984.

6,894   8,469   6,992   7,002   6,986   5,990

(10) Circle the numbers that are less than 3,275.

3,276   2,975   3,752   3,209   4,000   3,269

Ready for something a little different? Good!

Telling Time

Date　　/　　/

Name

Level ★ ★

Score ____/100

1 What time is it? Write the time under each clock.

5 points per question

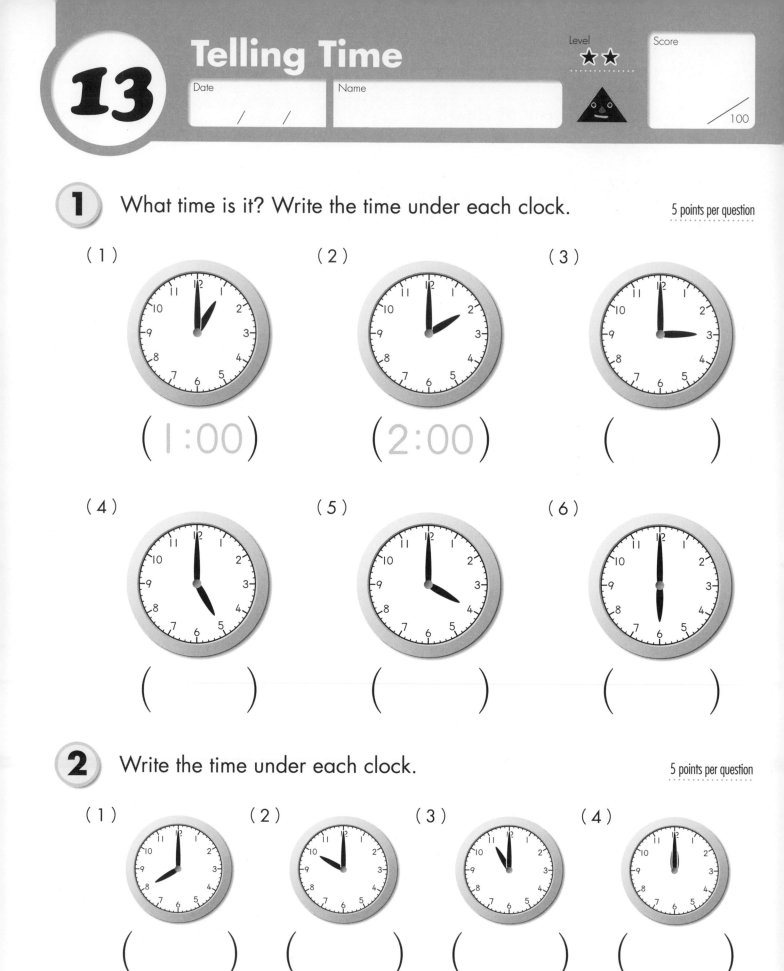

(1)

(1:00)

(2)

(2:00)

(3)

(　　　　)

(4)

(　　　　)

(5)

(　　　　)

(6)

(　　　　)

2 Write the time under each clock.

5 points per question

(1)

(　　　　)

(2)

(　　　　)

(3)

(　　　　)

(4)

(　　　　)

3 Write the time under each clock.   5 points per question

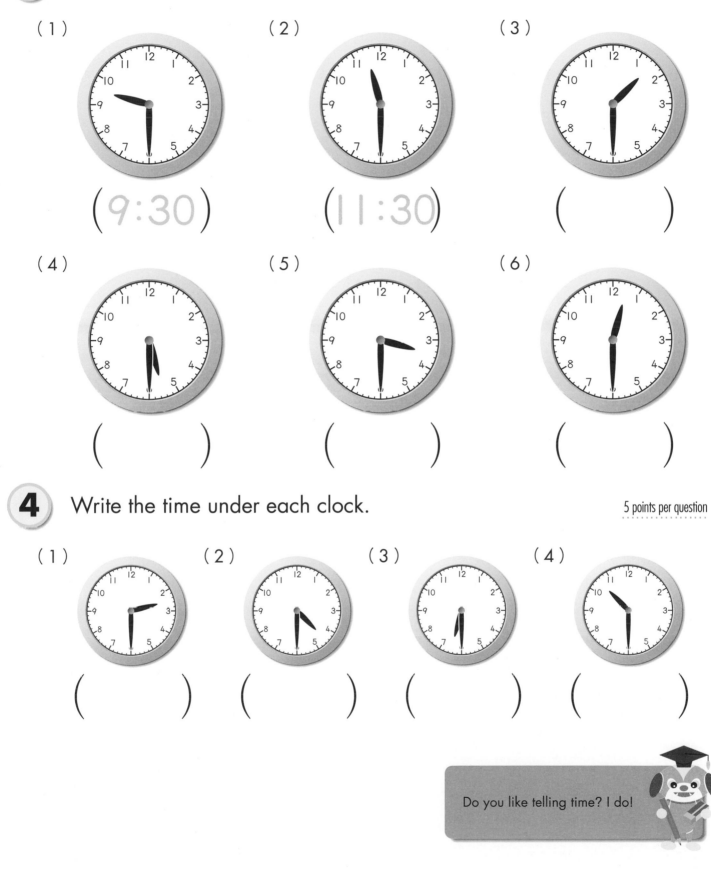

(1)

(9 : 30)

(2)

(11 : 30)

(3)

(   )

(4)

(   )

(5)

(   )

(6)

(   )

4 Write the time under each clock.   5 points per question

(1)

(   )

(2)

(   )

(3)

(   )

(4)

(   )

Do you like telling time? I do!

1 Write the time under each clock.

4 points per question

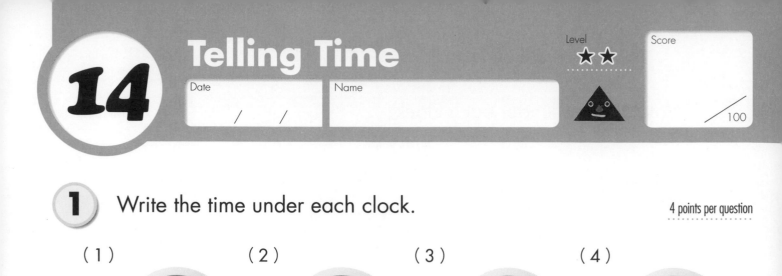

(1)   (2)   (3)   (4)

(8:05)   (   )   (   )   (   )

(5)   (6)   (7)   (8)

(8:25)   (8:30)   (   )   (   )

(9)   (10)   (11)   (12)

(   )   (   )   (   )   (   )

2　Write the time under each clock.

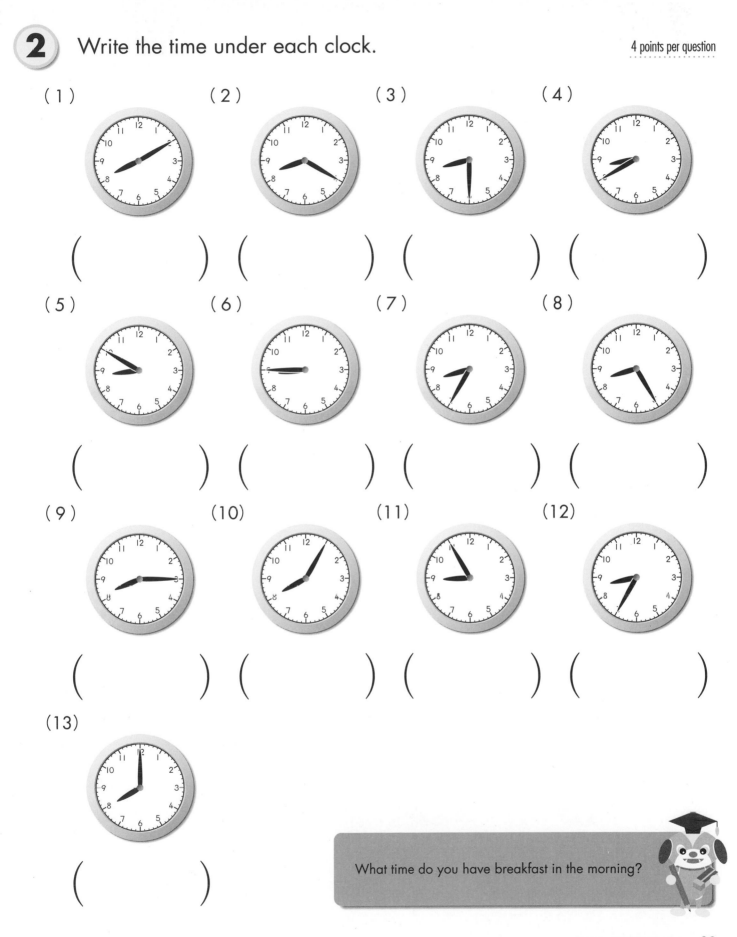

(1)

(　　　　　)

(2)

(　　　　　)

(3)

(　　　　　)

(4)

(　　　　　)

(5)

(　　　　　)

(6)

(　　　　　)

(7)

(　　　　　)

(8)

(　　　　　)

(9)

(　　　　　)

(10)

(　　　　　)

(11)

(　　　　　)

(12)

(　　　　　)

(13)

(　　　　　)

What time do you have breakfast in the morning?

15 Telling Time

Date   /   /

Name

Level ★ ★

Score   /100

1 Write the time under each clock.

5 points per question

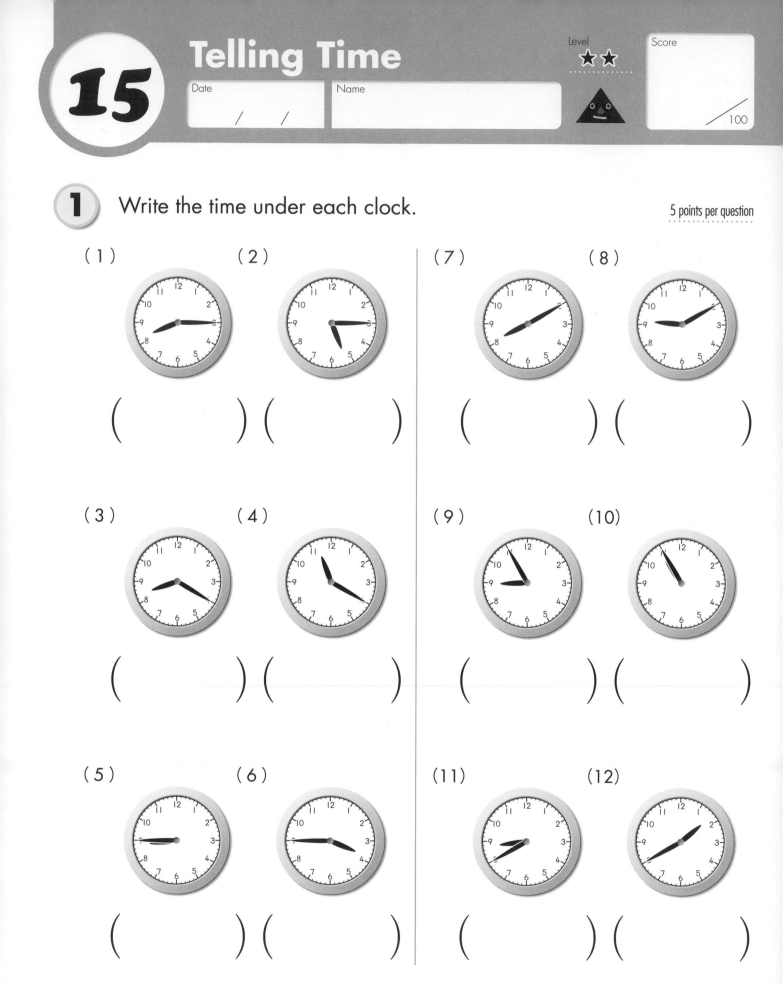

(1)   (2)

(   ) (   )

(3)   (4)

(   ) (   )

(5)   (6)

(   ) (   )

(7)   (8)

(   ) (   )

(9)   (10)

(   ) (   )

(11)   (12)

(   ) (   )

2 Write the time under each clock.

5 points per question

(1)

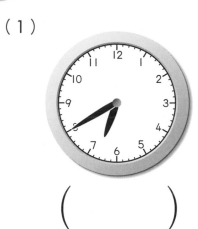

(   )

(2)

(   )

(3)

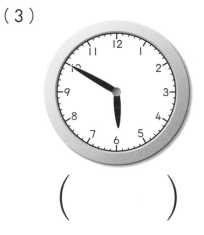

(   )

(4)

(   )

(5)

(   )

(6)

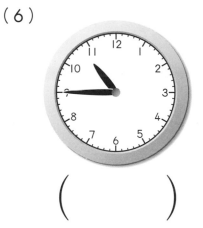

(   )

(7)

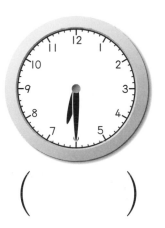

(   )

(8)

(   )

Now you're getting the hang of it!

31

16 Telling Time

Date   /   /

Name

Level
★★

Score
____/100

1 Draw the long hand on the face of each clock to match the time above.

5 points per question

(1) 9:00

(2) 9:30

(3) 11:00

(4) 11:30

(5) 12:00

(6) 12:30

2 Draw the long hand on the face of each clock to match the time above.

5 points per question

(1) 2:00

(2) 2:30

(3) 5:00

(4) 5:30

3 Draw the long hand on the face of each clock to match the time above.

5 points per question

(1) 9:05

(2) 9:10

(3) 9:20

(4) 9:40

(5) 11:40

(6) 11:15

4 Draw the long hand on the face of each clock to match the time above.

5 points per question

(1) 2:25

(2) 2:50

(3) 6:50

(4) 7:55

What time do you go to bed at night?

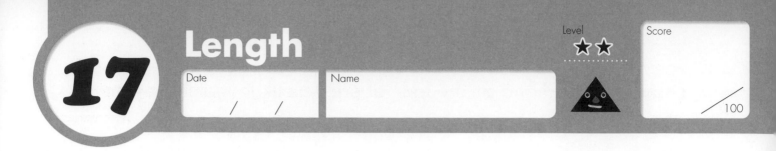

Length

1 Write a ✓ underneath the correct way to use the construction paper to measure the size of the notebook.

10 points for completion

ⓐ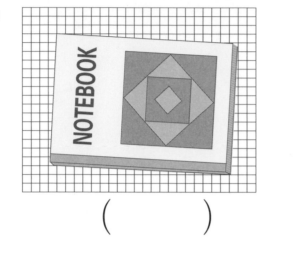

( )

ⓑ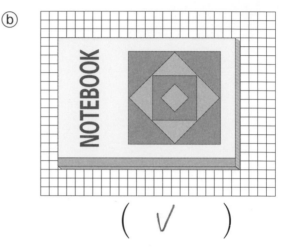

(✓)

2 Use the construction paper behind the book to answer questions about the length and width of this book. Each box on the paper is a unit.

10 points per question

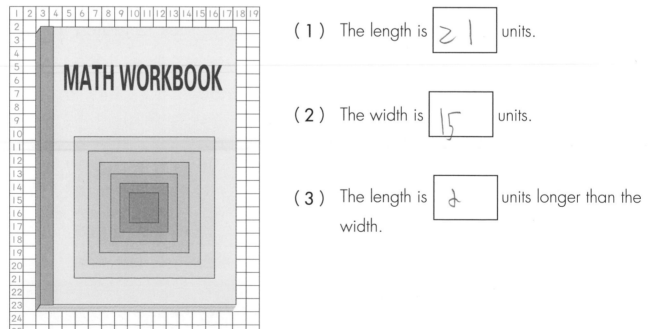

(1) The length is ⟨2 1⟩ units.

(2) The width is ⟨15⟩ units.

(3) The length is ⟨2⟩ units longer than the width.

34   © Kumon Publishing Co., Ltd.

3 One unit of the construction paper below is 1 inch (in.) long. What is the length of each object?

15 points per question

(1)

(tape)

(1 in.)

(2)

(eraser)

(2 in.)

(3)

(leaf)

(4 in.)

(4)

(pen)

(5 in.)

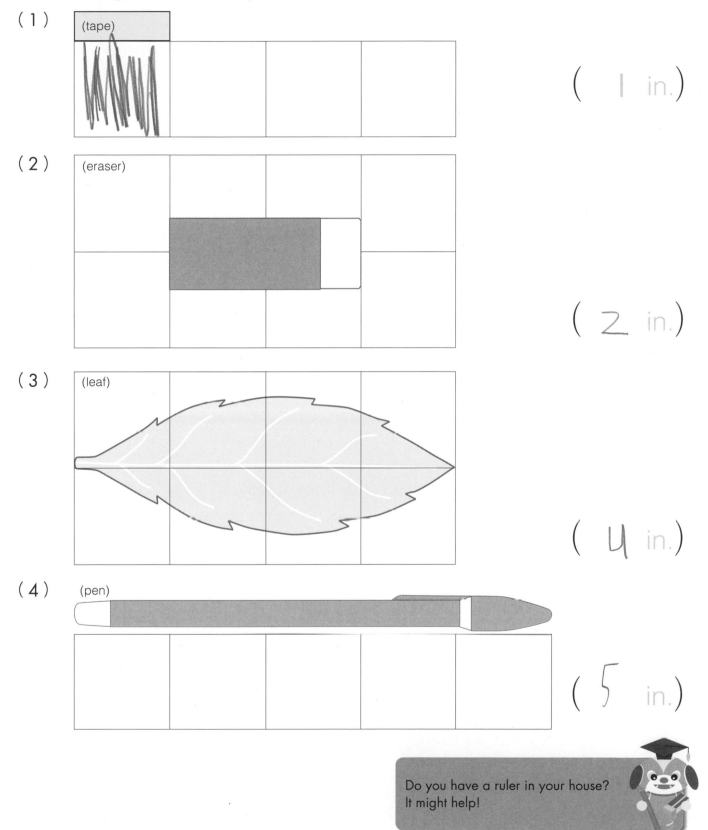

Do you have a ruler in your house? It might help!

Date / /

Name

Level ★★

Score /100

1 How many inches is it from the left side of the ruler to each box?

5 points per box

(1)

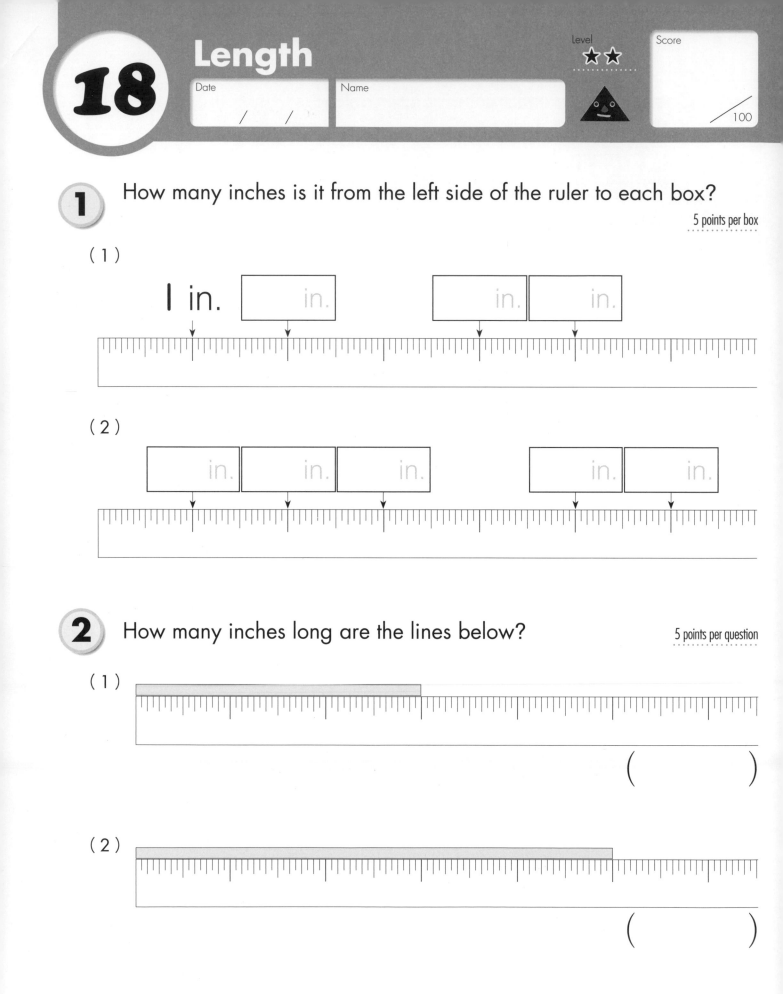

I in. | in. | in. | in.

(2)

in. | in. | in. | in. | in.

2 How many inches long are the lines below?

5 points per question

(1)

(   )

(2)

(   )

3 How many inches long are the pieces of tape below?   10 points per question

(1)

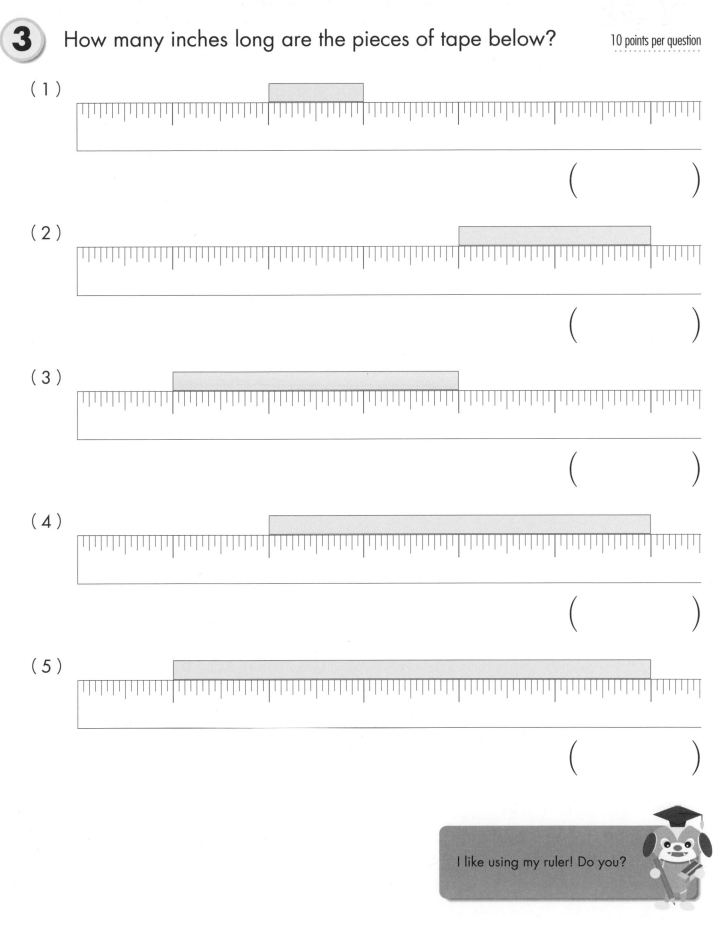

(   )

(2)

(   )

(3)

(   )

(4)

(   )

(5)

(   )

I like using my ruler! Do you?

37

Length

Date   /   /

Name

1 What is each length below?

4 points per question

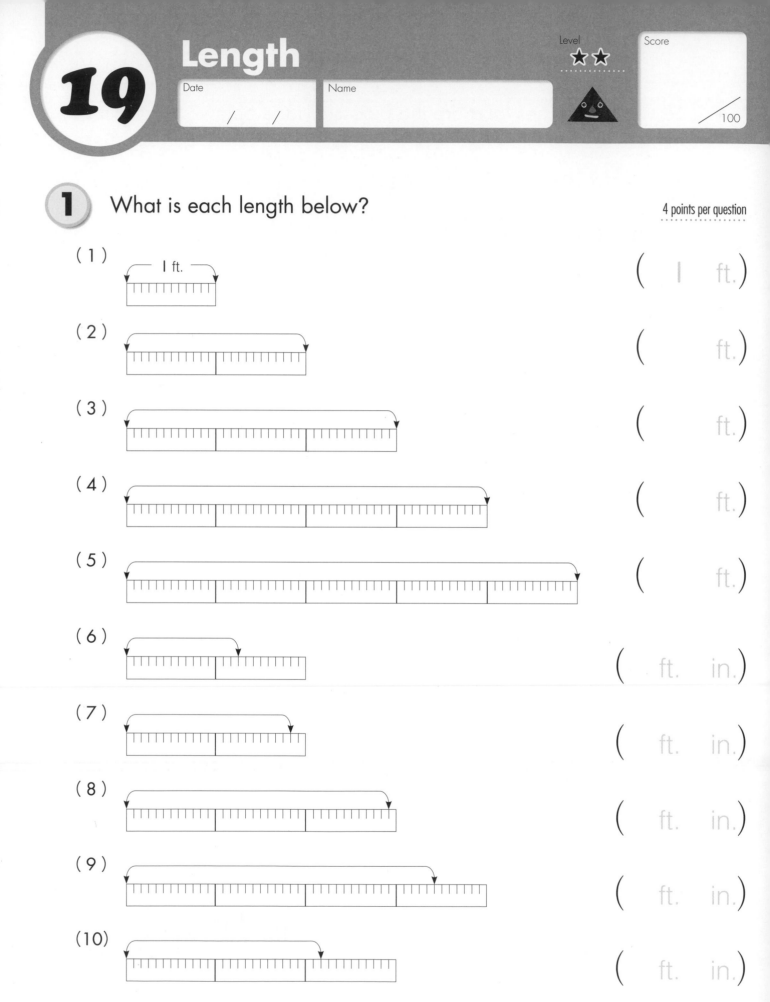

(1) 1 ft.   (1 ft.)

(2)   (ft.)

(3)   (ft.)

(4)   (ft.)

(5)   (ft.)

(6)   (ft. in.)

(7)   (ft. in.)

(8)   (ft. in.)

(9)   (ft. in.)

(10) (ft. in.)

Don't forget!

12 inches (in.) = 1 foot (ft.)

2 Write the appropriate number in each box.

5 points per question

(1) 1 ft. = ☐ in.

(2) 1 ft. 1 in. = ☐ in.

(3) 1 ft. 3 in. = ☐ in.

(4) 1 ft. 5 in. = ☐ in.

(5) 2 ft. = ☐ in.

(6) 2 ft. 2 in. = ☐ in.

(7) 12 in. = ☐ ft.

(8) 14 in. = ☐ ft. ☐ in.

(9) 16 in. = ☐ ft. ☐ in.

(10) 20 in. = ☐ ft. ☐ in.

(11) 25 in. = ☐ ft. ☐ in.

(12) 30 in. = ☐ ft. ☐ in.

Once you get the hang of it, it is not so tough!

39

20 Length

1 What is each length below?

4 points per question

(1) ⌐I yd.⌐

(**I** yd.)

(2)

(yd.)

(3)

(yd.)

(4)

(yd.)

(5)

(yd.)

(6)

(yd. ft.)

(7)

(yd. ft.)

(8)

(yd. ft.)

(9)

(yd. ft.)

(10)

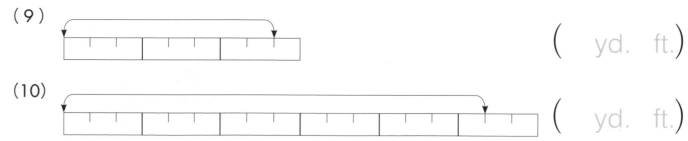

(yd. ft.)

Don't forget!

1 yard (yd.) = 3 feet (ft.)

2 Write the appropriate number in each box.

5 points per question

(1) 1 yd.= $\boxed{3}$ ft.

(2) 1 yd. 1 ft.= $\boxed{}$ ft.

(3) 2 yd.= $\boxed{}$ ft.

(4) 2 yd. 2 ft.= $\boxed{}$ ft.

(5) 3 yd. 1 ft.= $\boxed{}$ ft.

(6) 4 yd. 2 ft.= $\boxed{}$ ft.

(7) 3 ft.= $\boxed{}$ yd.

(8) 5 ft.= $\boxed{}$ yd. $\boxed{}$ ft.

(9) 7 ft.= $\boxed{}$ yd. $\boxed{}$ ft.

(10) 10 ft.= $\boxed{}$ yd. $\boxed{}$ ft.

(11) 12 ft.= $\boxed{}$ yd.

(12) 16 ft.= $\boxed{}$ yd. $\boxed{}$ ft.

Inches, then feet, then yards!
We are getting longer with every page.

1 You want to measure the objects below. Which tool is the best for each measurement? Select the appropriate tool and write its letter next to each object.

4 points per question

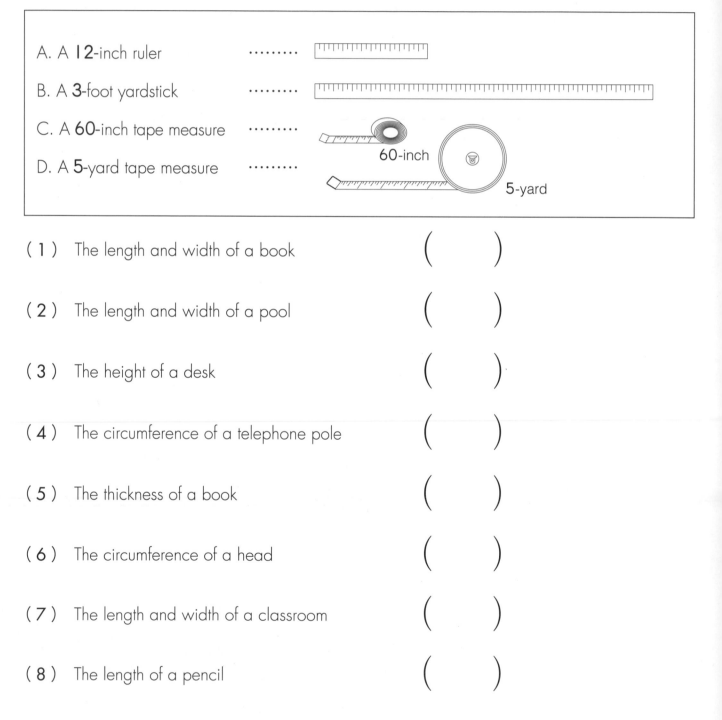

A. A **12**-inch ruler

B. A **3**-foot yardstick

C. A **60**-inch tape measure ... 60-inch

D. A **5**-yard tape measure ... 5-yard

(1) The length and width of a book (   )

(2) The length and width of a pool (   )

(3) The height of a desk (   )

(4) The circumference of a telephone pole (   )

(5) The thickness of a book (   )

(6) The circumference of a head (   )

(7) The length and width of a classroom (   )

(8) The length of a pencil (   )

2 Read each tape measure and write the appropriate length in each box.

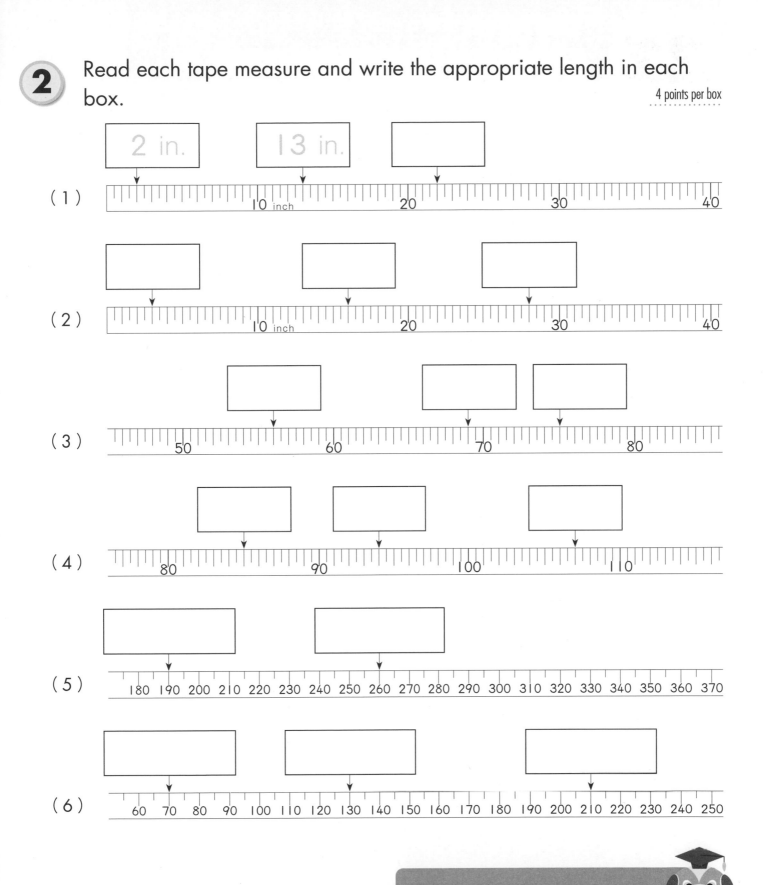

(1) 2 in.　13 in.

How big is your room? Try to measure it!

22 Length

Date / /

Name

Level ★★

Score / 100

1 How many centimeters is it from the left side of the ruler to each box?

5 points per box

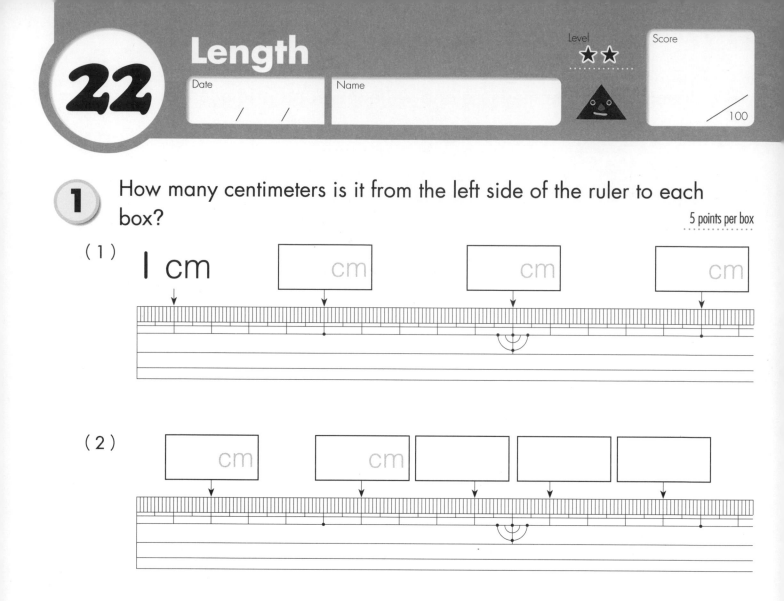

(1)

1 cm   cm   cm   cm

(2)

cm   cm

2 How many centimeters long is each line below?

5 points per question

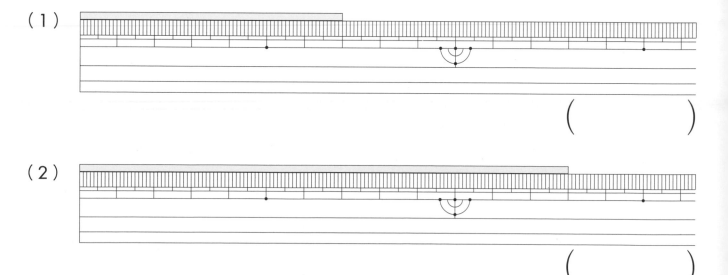

(1)

(   )

(2)

(   )

3 How many centimeters long is each piece of tape below? 10 points per question

(1)

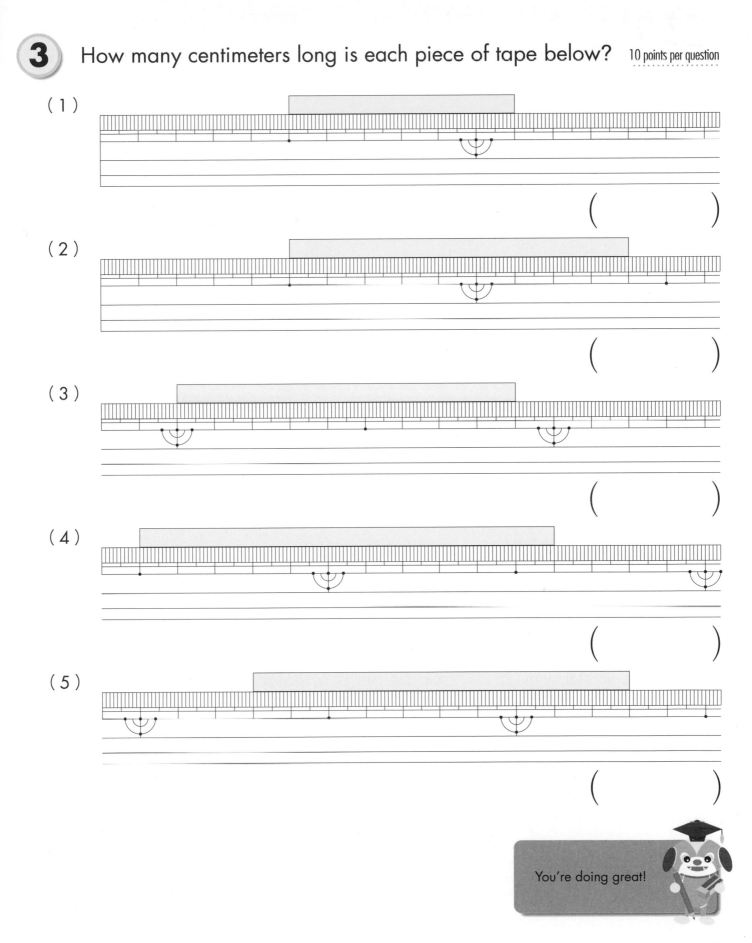

( )

(2)

( )

(3)

( )

(4)

( )

(5)

( )

You're doing great!

23 Length

Date

Name

Level ★★

Score

/100

1 Each ruler below is 30 centimeters long. How many centimeters long is each distance marked by the arrows below?

5 points per question

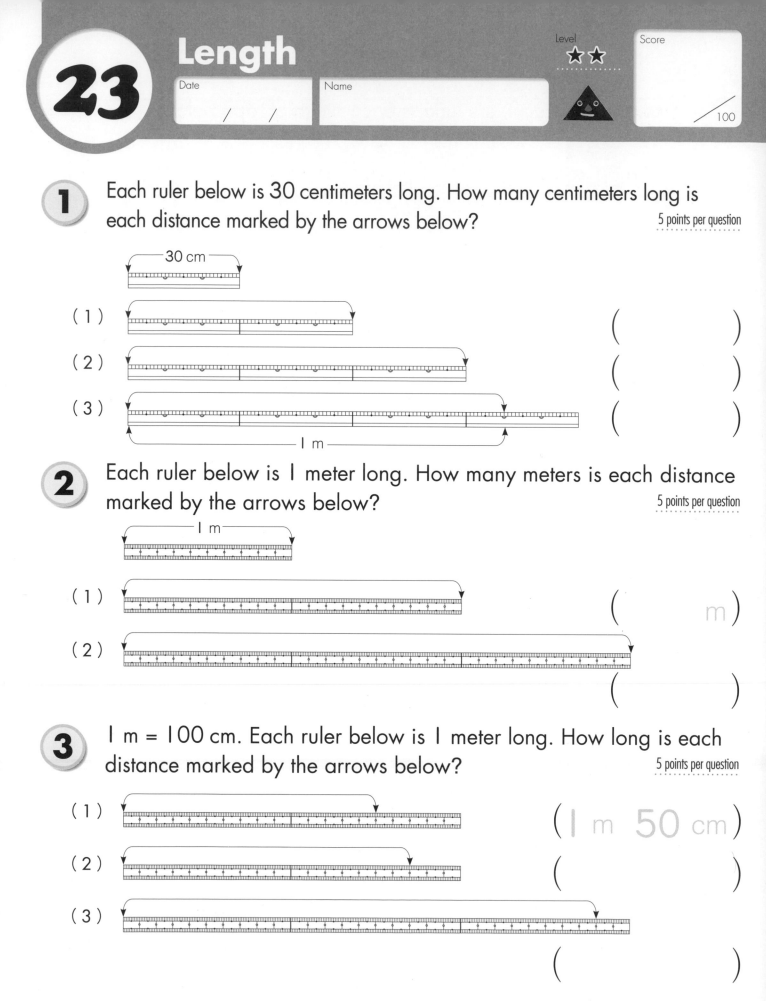

30 cm

(1) (   )

(2) (   )

(3) (   )

1 m

2 Each ruler below is 1 meter long. How many meters is each distance marked by the arrows below?

5 points per question

1 m

(1) (m)

(2) (   )

3 1 m = 100 cm. Each ruler below is 1 meter long. How long is each distance marked by the arrows below?

5 points per question

(1) (1 m 50 cm)

(2) (   )

(3) (   )

4 Write the appropriate number in each box.

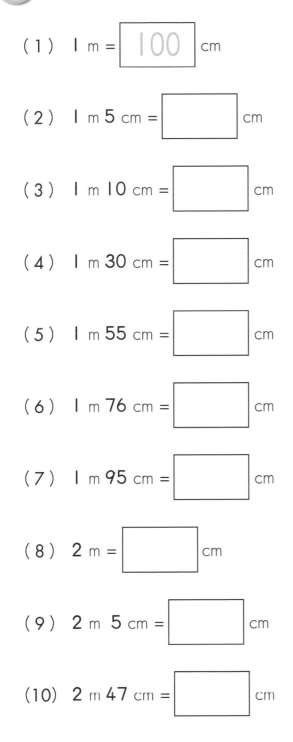

(1) 1 m = ⬜ 100 ⬜ cm

(2) 1 m 5 cm = ⬜ cm

(3) 1 m 10 cm = ⬜ cm

(4) 1 m 30 cm = ⬜ cm

(5) 1 m 55 cm = ⬜ cm

(6) 1 m 76 cm = ⬜ cm

(7) 1 m 95 cm = ⬜ cm

(8) 2 m = ⬜ cm

(9) 2 m 5 cm = ⬜ cm

(10) 2 m 47 cm = ⬜ cm

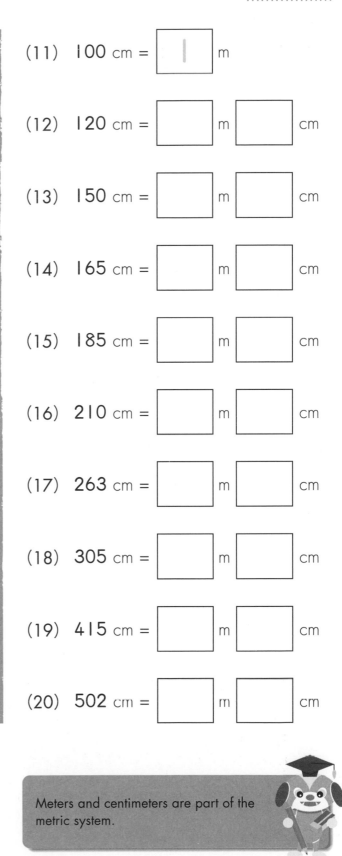

(11) 100 cm = ⬜ 1 ⬜ m

(12) 120 cm = ⬜ m ⬜ cm

(13) 150 cm = ⬜ m ⬜ cm

(14) 165 cm = ⬜ m ⬜ cm

(15) 185 cm = ⬜ m ⬜ cm

(16) 210 cm = ⬜ m ⬜ cm

(17) 263 cm = ⬜ m ⬜ cm

(18) 305 cm = ⬜ m ⬜ cm

(19) 415 cm = ⬜ m ⬜ cm

(20) 502 cm = ⬜ m ⬜ cm

Meters and centimeters are part of the metric system.

24 Weight

1 Read the weight on each scale and write it below.   5 points per question

(1)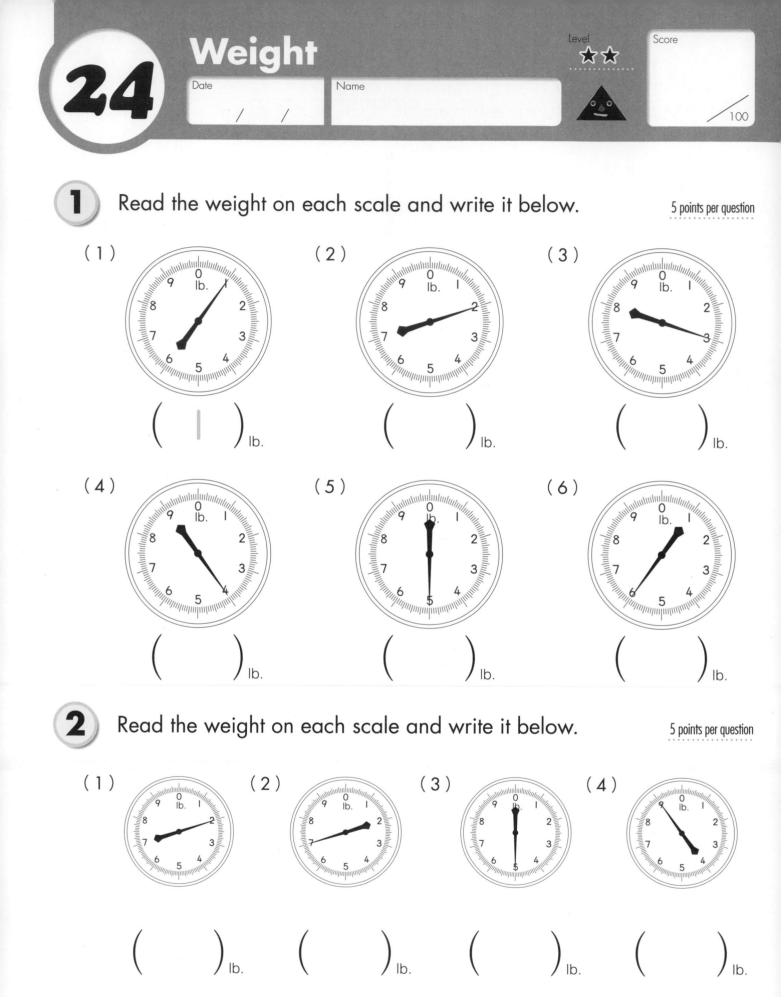

(|) lb.

(2)

(   ) lb.

(3)

(   ) lb.

(4)

(   ) lb.

(5)

(   ) lb.

(6)

(   ) lb.

2 Read the weight on each scale and write it below.   5 points per question

(1)

(   ) lb.

(2)

(   ) lb.

(3)

(   ) lb.

(4)

(   ) lb.

3 Read the weight on each scale and write it below.　　　5 points per question

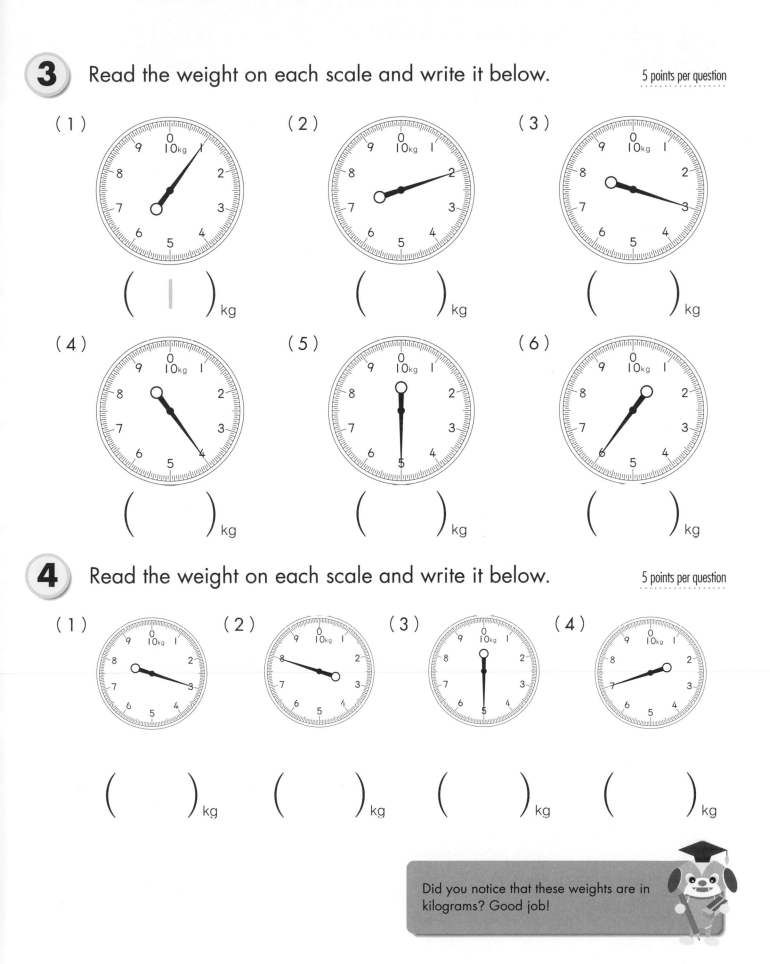

(1)　　　　　　　　　　　(2)　　　　　　　　　　　(3)

(　1　) kg　　　　　(　　　) kg　　　　　(　　　) kg

(4)　　　　　　　　　　　(5)　　　　　　　　　　　(6)

(　　　) kg　　　　　(　　　) kg　　　　　(　　　) kg

4 Read the weight on each scale and write it below.　　　5 points per question

(1)　　　　　　(2)　　　　　　(3)　　　　　　(4)

(　　　) kg　　(　　　) kg　　(　　　) kg　　(　　　) kg

Did you notice that these weights are in kilograms? Good job!

1 These weights are in customary units (pounds). Read the weight on each scale and write it below.

6 points per question

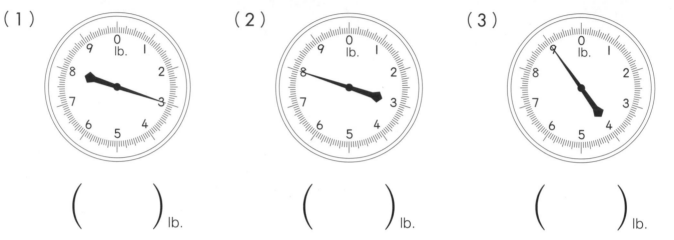

(1)

( ) lb.

(2)

( ) lb.

(3)

( ) lb.

2 These weights are in metric units (kilograms). Read the weight on each scale and write it below.

6 points per question

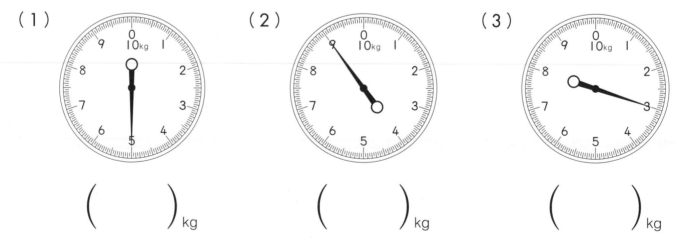

(1)

( ) kg

(2)

( ) kg

(3)

( ) kg

Read the weight on each scale and write it below.

8 points per question

(1) (　　　　) lb.

(2) (　　　　) kg

(3) (　　　　) kg

(4) (　　　　) lb.

(5) (　　　　) lb.

(6) (　　　　) kg

(7) (　　　　) kg

(8) (　　　　) lb.

Did you notice the different units in each question? Good job!

Counting Coins

1 Add the value of each group of coins. Then write the amount on the right.

5 points per question

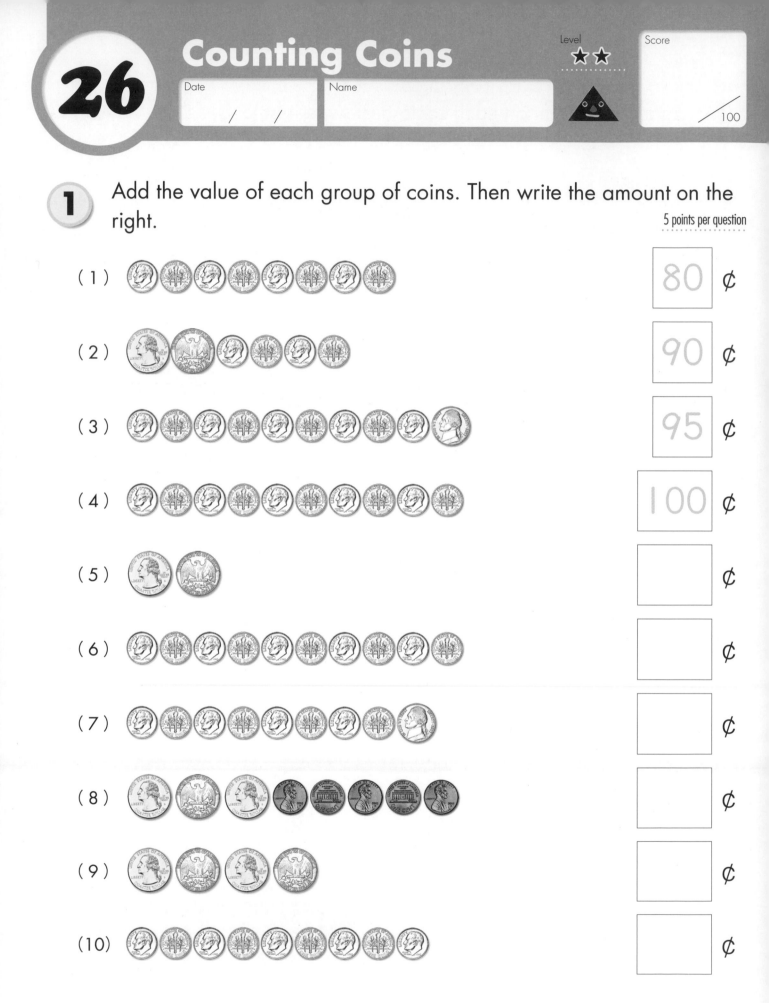

(1) 80 ¢

(2) 90 ¢

(3) 95 ¢

(4) 100 ¢

(5) ¢

(6) ¢

(7) ¢

(8) ¢

(9) ¢

(10) ¢

100 ¢ $= \$1$ (100 cents equals 1 dollar.)

2 Add the value of each row of coins. Then write the amount on the right.

5 points per question

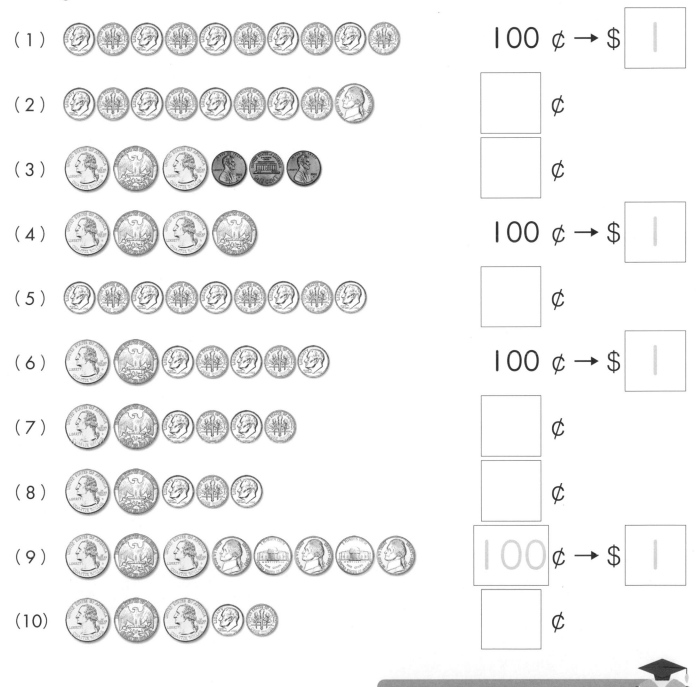

(1) 100 ¢ → $ \boxed{1}$

(2) $\boxed{}$ ¢

(3) $\boxed{}$ ¢

(4) 100 ¢ → $ \boxed{1}$

(5) $\boxed{}$ ¢

(6) 100 ¢ → $ \boxed{1}$

(7) $\boxed{}$ ¢

(8) $\boxed{}$ ¢

(9) $\boxed{100}$ ¢ → $ \boxed{1}$

(10) $\boxed{}$ ¢

Counting money is tough. Well done!

1 Add the value of the money in each box. Then write the amount in the box on the right.

6 points per question

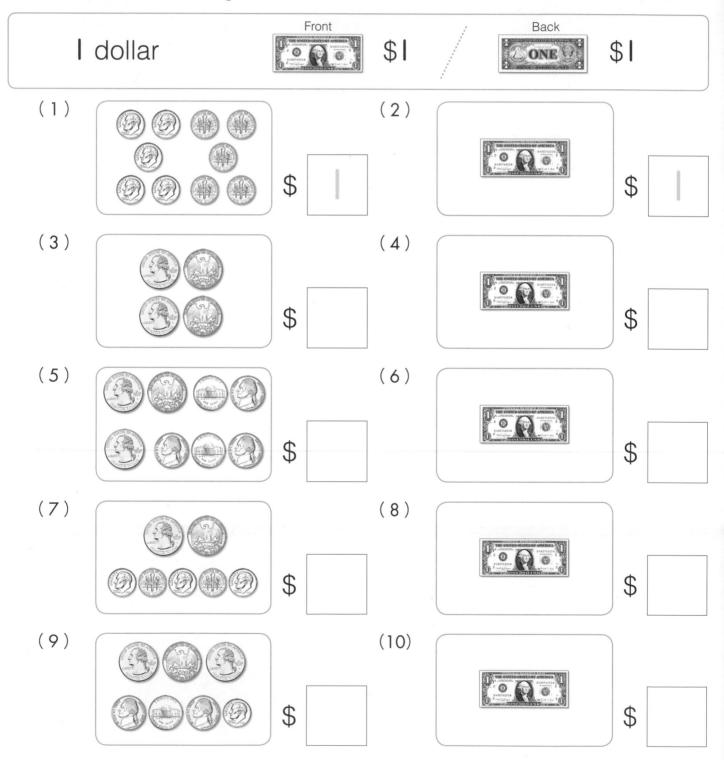

	Front		Back	
1 dollar		$1		$1

(1) $ |

(2) $ |

(3) $

(4) $

(5) $

(6) $

(7) $

(8) $

(9) $

(10) $

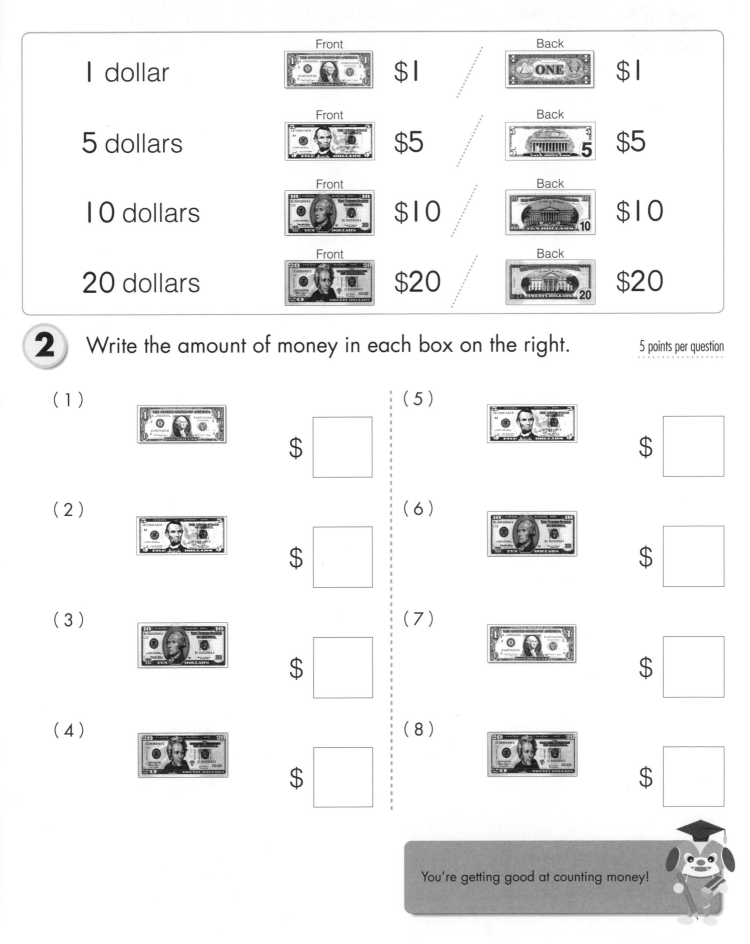

	Front		Back	
1 dollar		$1		$1
5 dollars		$5		$5
10 dollars		$10		$10
20 dollars		$20		$20

2 Write the amount of money in each box on the right.

5 points per question

(1) $ ☐

(2) $ ☐

(3) $ ☐

(4) $ ☐

(5) $ ☐

(6) $ ☐

(7) $ ☐

(8) $ ☐

You're getting good at counting money!

28 Counting Money

Level ★★★

Score
/100

Date / /

Name

1 Count the money in each row. Then write the amount on the right.

5 points per question

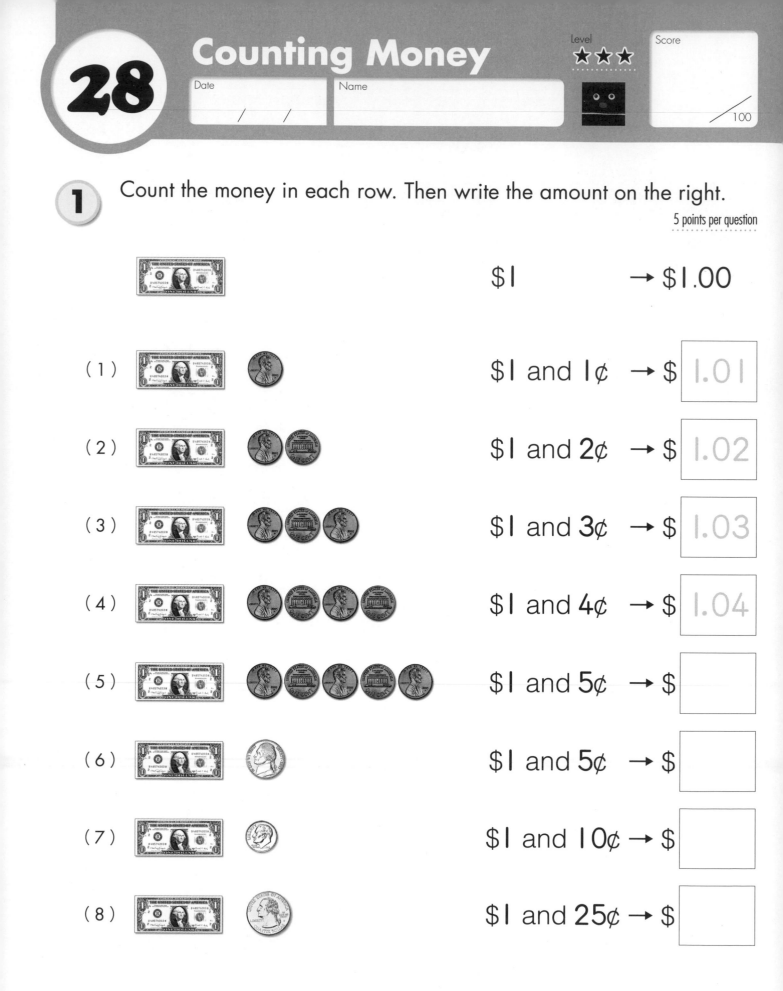

$1 → $1.00

(1) $1 and 1¢ → $ 1.01

(2) $1 and 2¢ → $ 1.02

(3) $1 and 3¢ → $ 1.03

(4) $1 and 4¢ → $ 1.04

(5) $1 and 5¢ → $

(6) $1 and 5¢ → $

(7) $1 and 10¢ → $

(8) $1 and 25¢ → $

2 Count the money in each row. Then write the amount on the right.

6 points per question

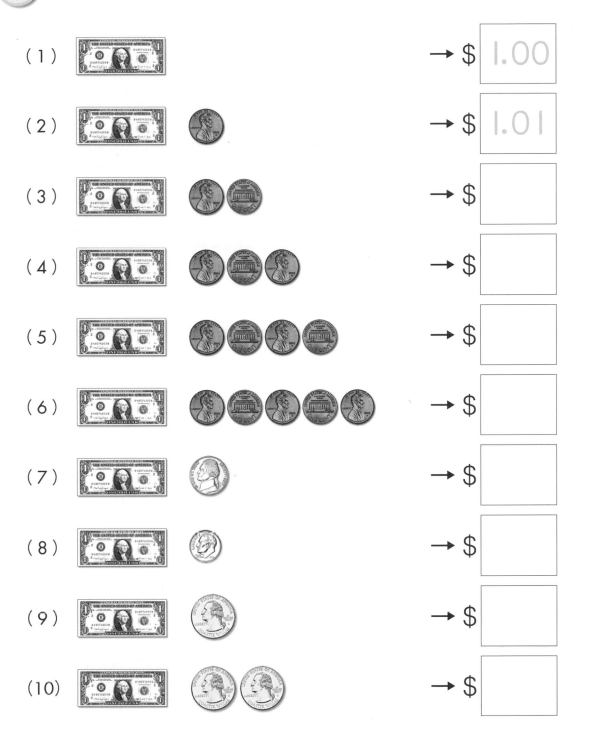

(1) → $ 1.00

(2) → $ 1.01

(3) → $

(4) → $

(5) → $

(6) → $

(7) → $

(8) → $

(9) → $

(10) → $

Do you have a piggybank?
How much do you think is in it?

Exchange Unit

Level ★★

Score /100

1 Add the value of each row of coins. Then write or trace the amount on the right.

5 points per question

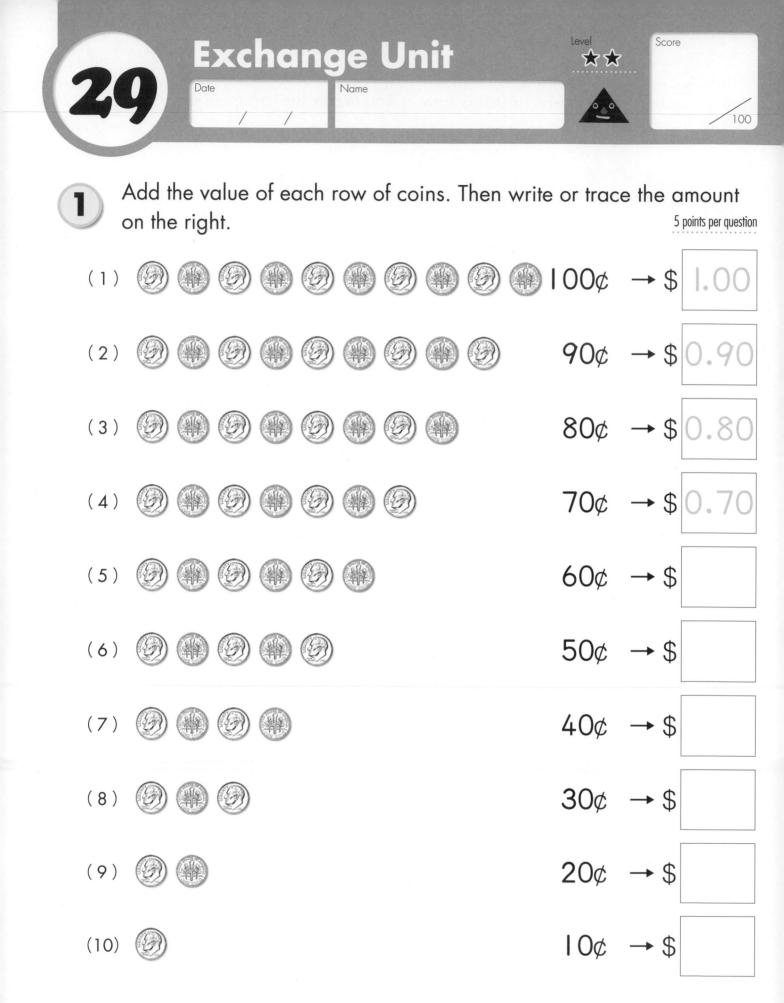

(1) 100¢ → $ 1.00

(2) 90¢ → $ 0.90

(3) 80¢ → $ 0.80

(4) 70¢ → $ 0.70

(5) 60¢ → $

(6) 50¢ → $

(7) 40¢ → $

(8) 30¢ → $

(9) 20¢ → $

(10) 10¢ → $

2 Add the value of each row of coins. Then write the amount on the right.

5 points per question

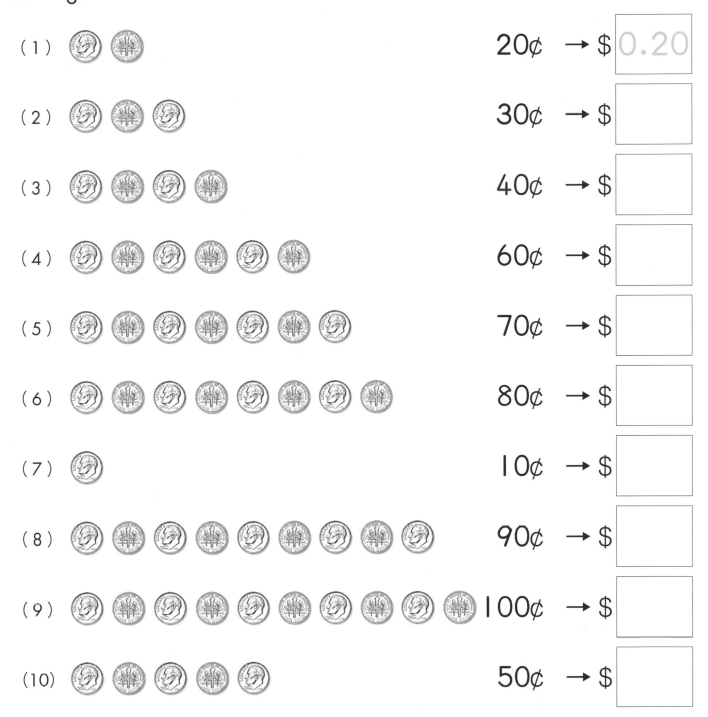

(1) 20¢ → $ 0.20

(2) 30¢ → $

(3) 40¢ → $

(4) 60¢ → $

(5) 70¢ → $

(6) 80¢ → $

(7) 10¢ → $

(8) 90¢ → $

(9) 100¢ → $

(10) 50¢ → $

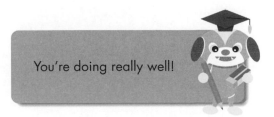

You're doing really well!

30 Money

Level ☆☆

Score /100

Date / / Name

1 Count the money in each row and write the amount on the right. Then circle the amount that is larger.

10 points per question

(1)
ⓐ 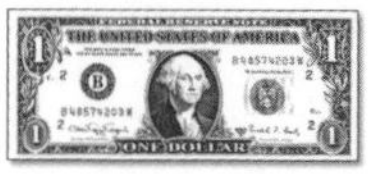 $ 1.00
ⓑ 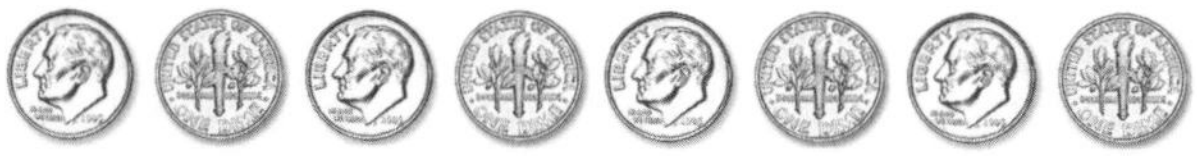 $ 0.80

(2)
ⓐ 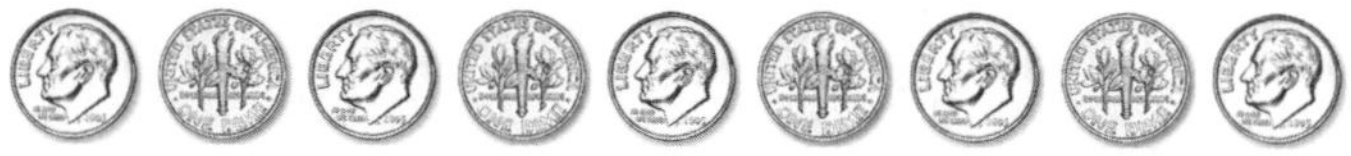 $
ⓑ 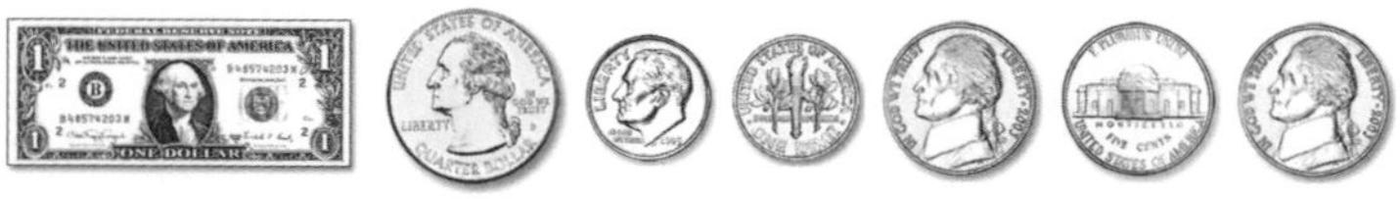 $

(3)
ⓐ 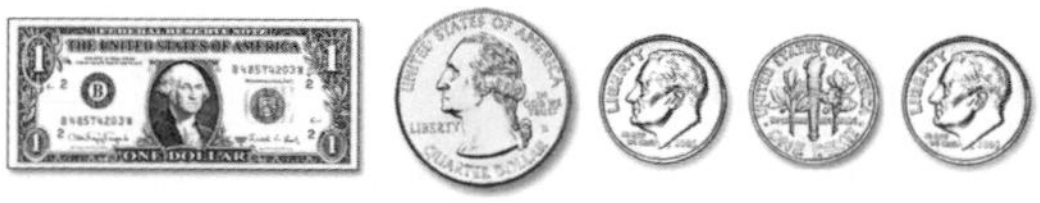 $
ⓑ 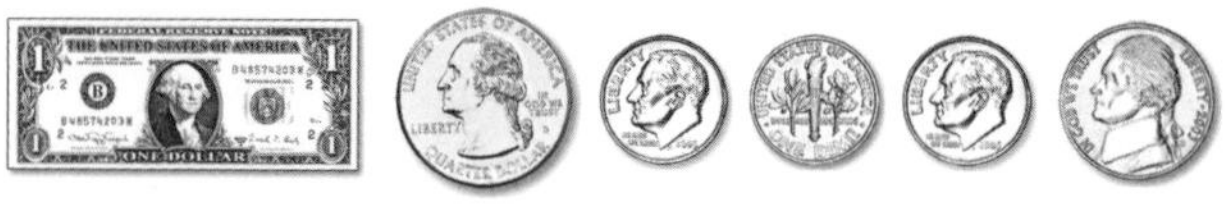 $

(4)
ⓐ 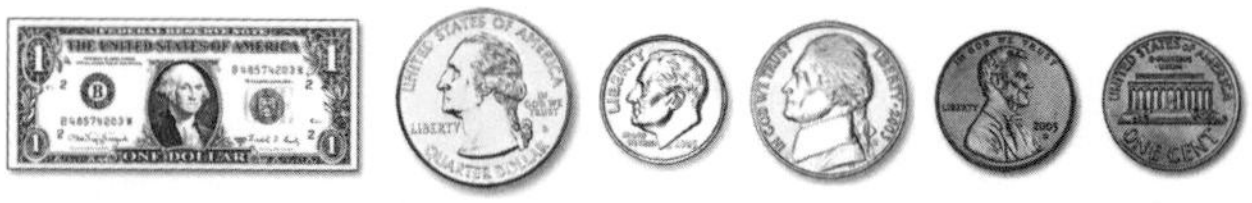 $
ⓑ 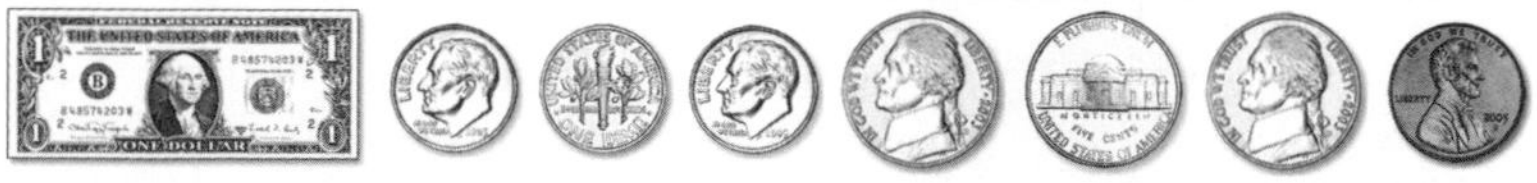 $

(5)
ⓐ 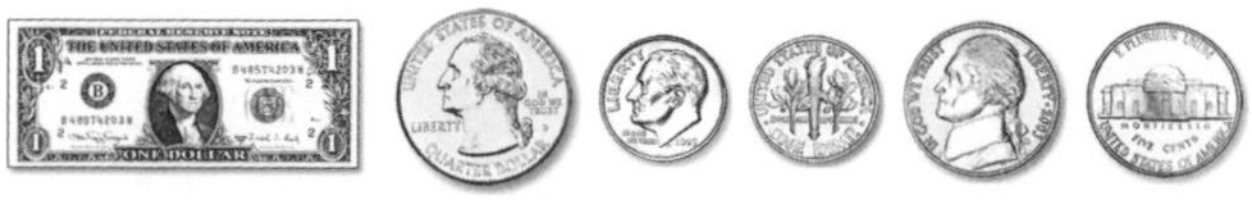 $
ⓑ 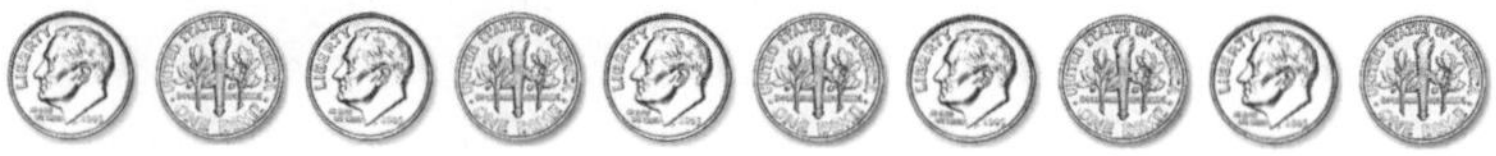 $

2 Count the money in each row and write the amount on the right.
Then circle the amount that is larger.

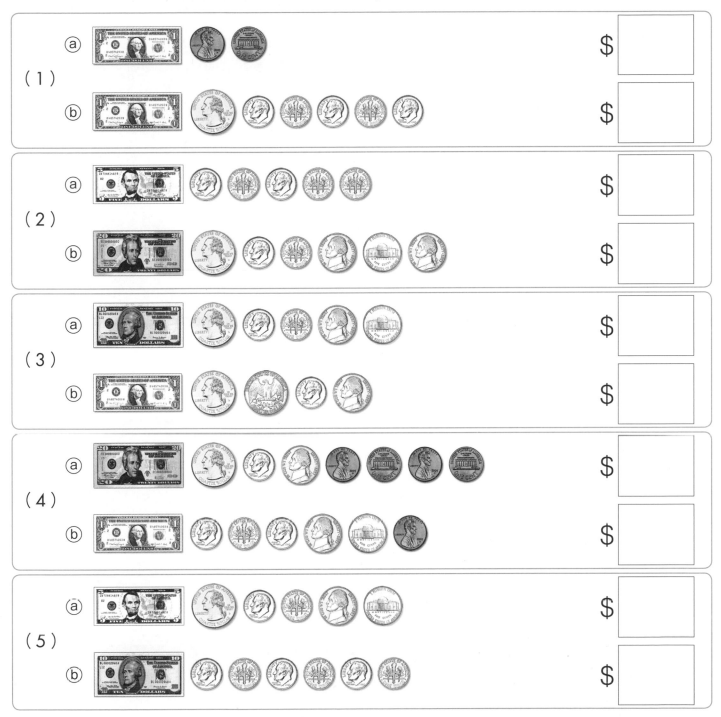

(1)
ⓐ $
ⓑ $

(2)
ⓐ $
ⓑ $

(3)
ⓐ $
ⓑ $

(4)
ⓐ $
ⓑ $

(5)
ⓐ $
ⓑ $

Are you ready for something different?
Good!

Triangles & Quadrilaterals

Level ★★

Date   /   /

Name

Score   /100

1 How many triangles are used to create the figures below?   5 points per question

(1)

(   )

(2)

(   )

(3)

(   )

2 Use triangles to create the figures below.   5 points per question

● Use two triangles here. Draw a line to separate the two triangles in each shape.

(1)

(2)

(3)

● Use three triangles here. Draw two lines to separate the three triangles in each shape.

(4)

(5)

(6)

(7)

3 How many triangles are in the shapes below?

(1)

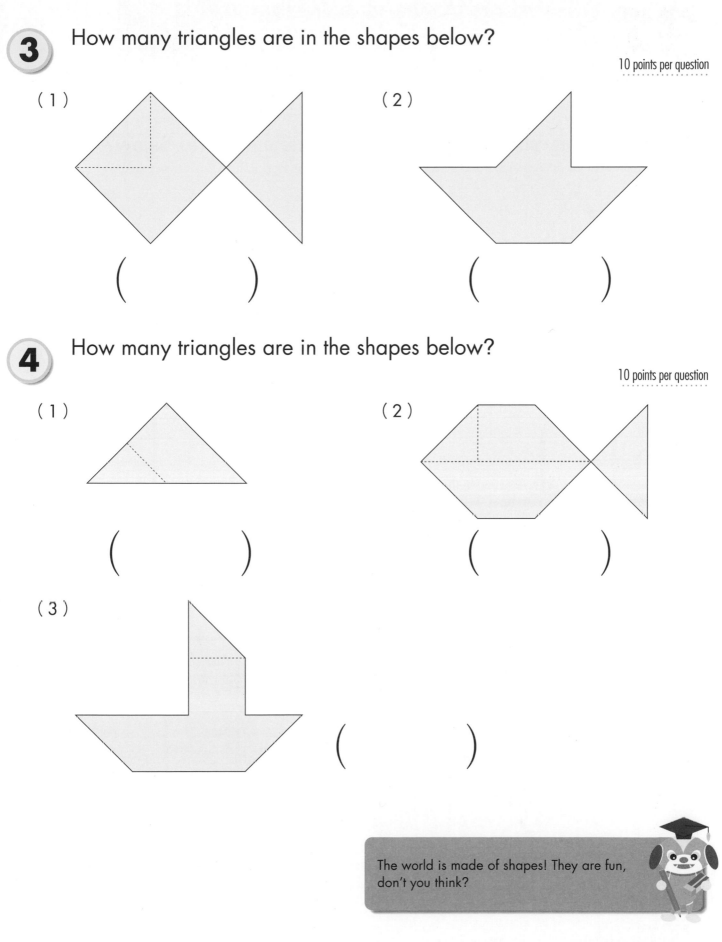

(   )

(2)

(   )

4 How many triangles are in the shapes below?

10 points per question

(1)

(   )

(2)

(   )

(3)

(   )

The world is made of shapes! They are fun, don't you think?

1 How many sticks do you need to create the figures below? 5 points per question

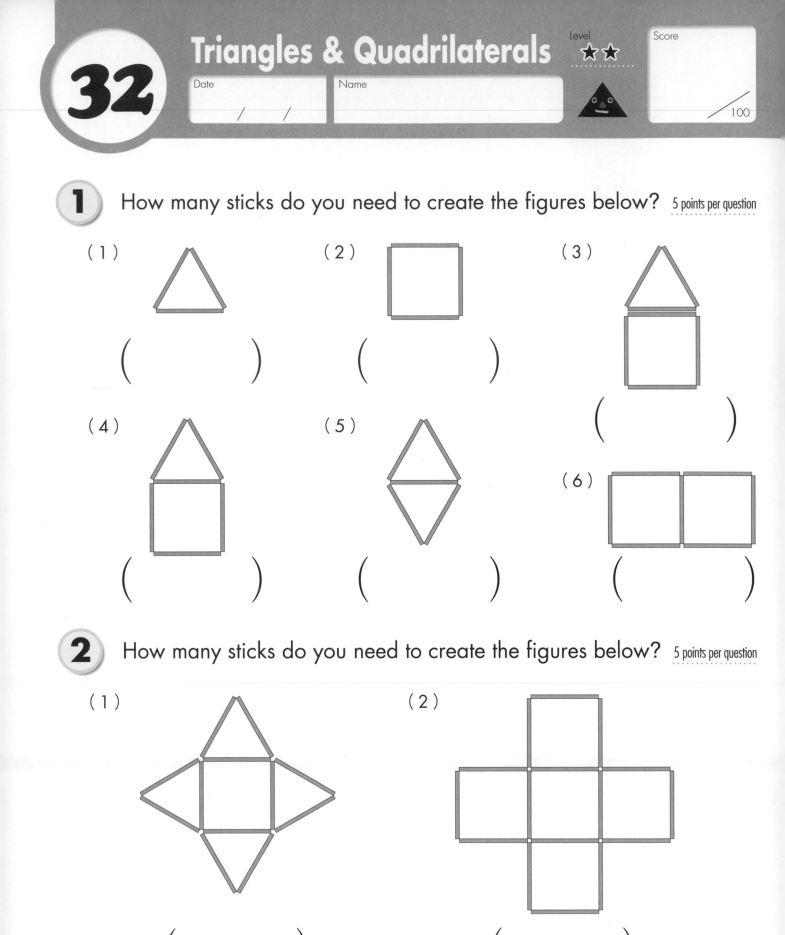

(1)

( )

(2)

( )

(3)

( )

(4)

( )

(5)

( )

(6)

( )

2 How many sticks do you need to create the figures below? 5 points per question

(1)

( )

(2)

( )

3 On the graph paper pictured here, you drew some shapes by connecting points with straight lines. Sort the shapes you made into the two groups below.

15 points per question

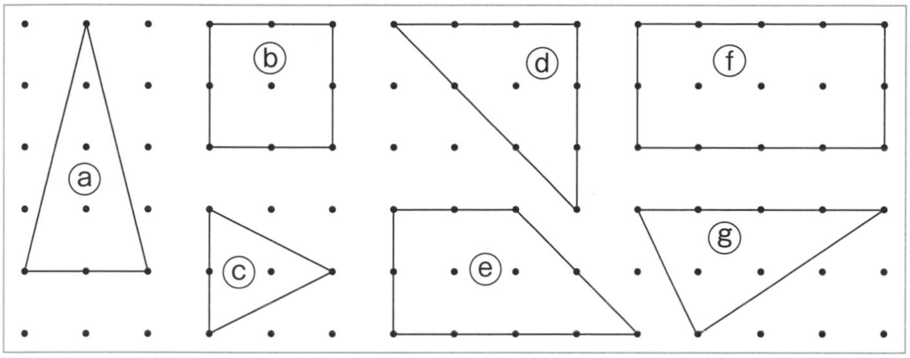

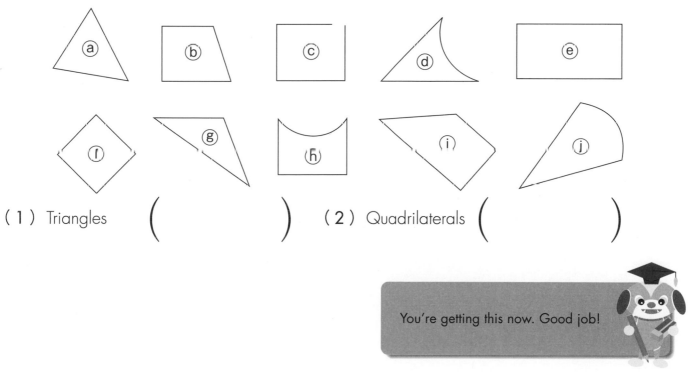

(1 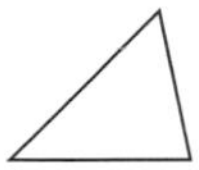) ◺ Triangles (shapes created with three straight line segments)

(   )

(2 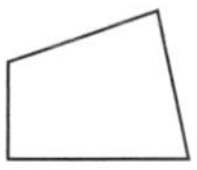) ▱ Quadrilaterals (shapes created with four straight line segments)

(   )

4 Find all the triangles and quadrilaterals in the shapes pictured below. Use the letter for each shape to answer the question.

15 points per question

(1) Triangles (   ) (2) Quadrilaterals (   )

You're getting this now. Good job!

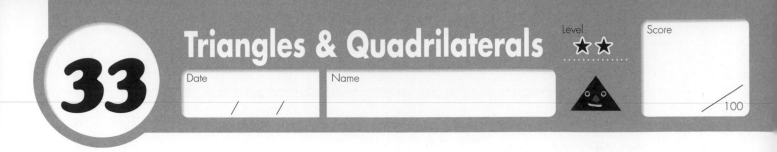

Triangles & Quadrilaterals

Level ★★

Date / /

Name

Score / 100

1 Finish drawing the triangles below by connecting the three points with straight lines.

8 points per question

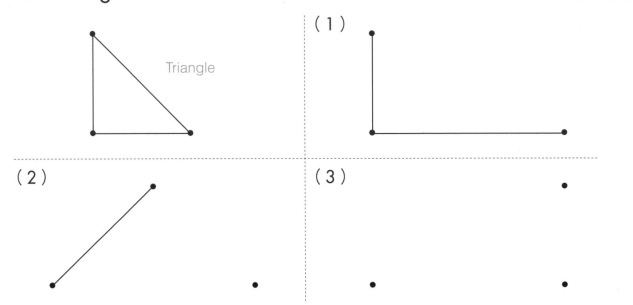

Triangle

(1)

(2)

(3)

2 Finish drawing the quadrilaterals below by connecting the four points with straight lines.

8 points per question

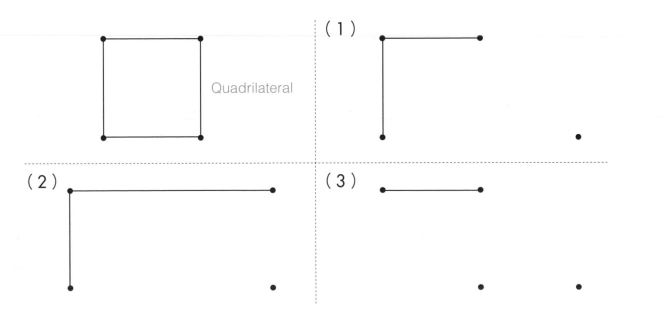

Quadrilateral

(1)

(2)

(3)

3 Draw 3 triangles by connecting points below.

Triangle

4 Draw 3 quadrilaterals by connecting points below.

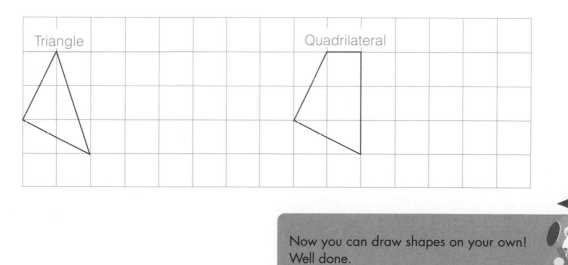

Quadrilateral

5 Draw a triangle and a quadrilateral on the grid below.

Triangle   Quadrilateral

Now you can draw shapes on your own!
Well done.

34 Triangles & Quadrilaterals

Level ★★

Score /100

Date / /

Name

1 You cut a triangle into two pieces by cutting along the dashed line that passes through the vertex. What kind of shapes did you make? — 10 points

(Triangles)

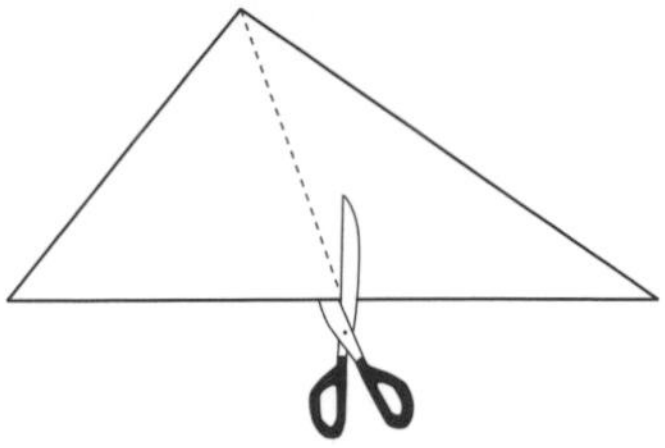

2 You cut a triangle into two pieces by cutting along the dashed line pictured to the right. What kind of shapes did you make? — 10 points

(   ) and (   )

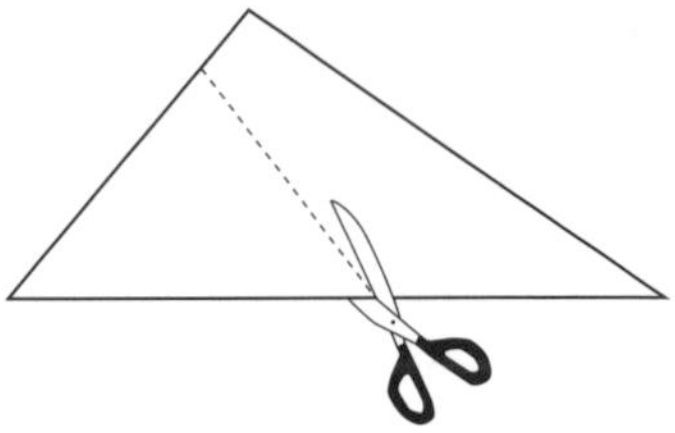

3 You cut these triangles along the dashed lines pictured here. Use each triangle's letter to answer the questions below. — 10 points per question

ⓐ

ⓑ

ⓒ

ⓓ

ⓔ

ⓕ

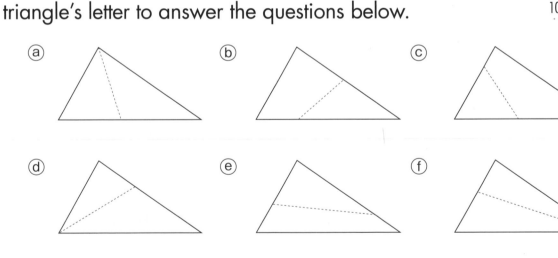

(1) Which dashed lines create two triangles?

(   )

(2) Which dashed lines create a triangle and a quadrilateral?

(   )

4 You cut the quadrilaterals into two pieces as shown here. What kind of shapes did you make in each case?

10 points per question

(1)

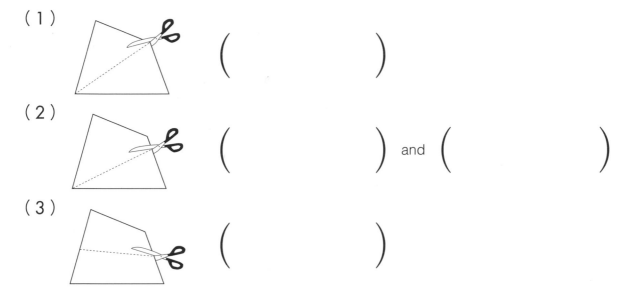

(   )

(2)

(   ) and (   )

(3)

(   )

5 You cut these quadrilaterals along the dashed lines pictured here. Use each quadrilateral's letter to answer the questions below.

10 points per question

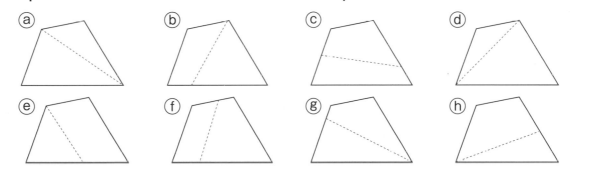

ⓐ ⓑ ⓒ ⓓ

ⓔ ⓕ ⓖ ⓗ

(1) Which dashed lines create two triangles?

(   )

(2) Which dashed lines create a triangle and a quadrilateral?

(   )

(3) Which dashed lines create two quadrilaterals?

(   )

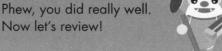

Phew, you did really well.
Now let's review!

35 Review

Level ★★★

Score

/100

Date   /   /

Name

1 Count the objects below and write the amount in each box.

6 points per question

(1)

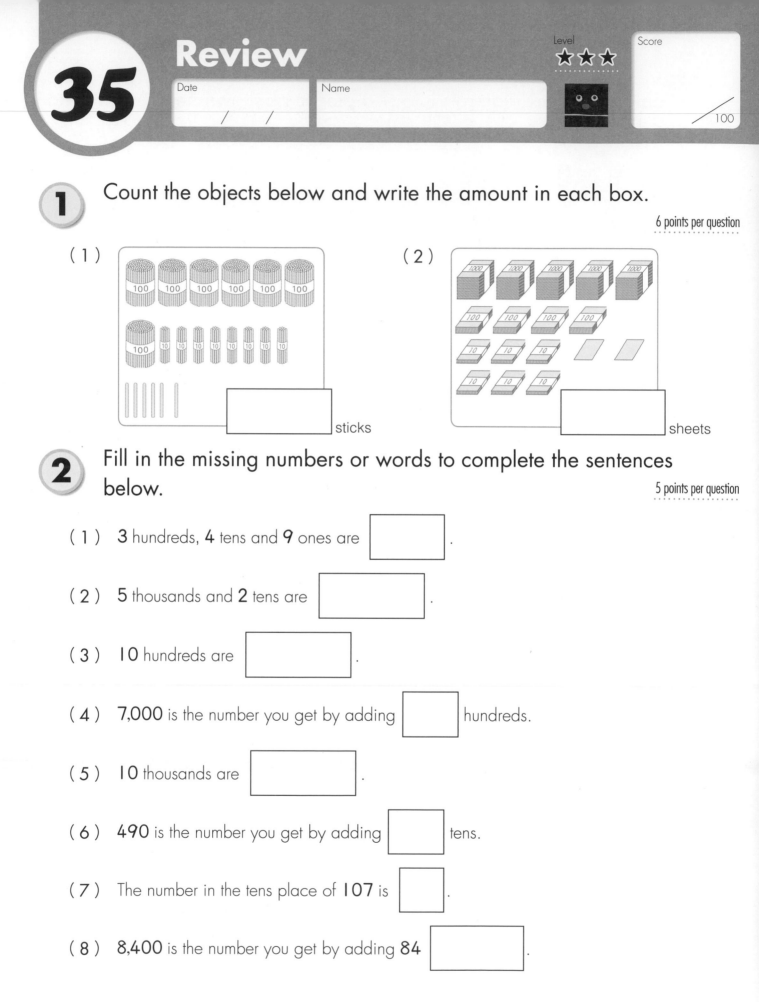

sticks

(2)

sheets

2 Fill in the missing numbers or words to complete the sentences below.

5 points per question

(1)   3 hundreds, 4 tens and 9 ones are ☐ .

(2)   5 thousands and 2 tens are ☐ .

(3)   10 hundreds are ☐ .

(4)   7,000 is the number you get by adding ☐ hundreds.

(5)   10 thousands are ☐ .

(6)   490 is the number you get by adding ☐ tens.

(7)   The number in the tens place of 107 is ☐ .

(8)   8,400 is the number you get by adding 84 ☐ .

3 What time is it? Write the time under each clock. 6 points per question

(1) 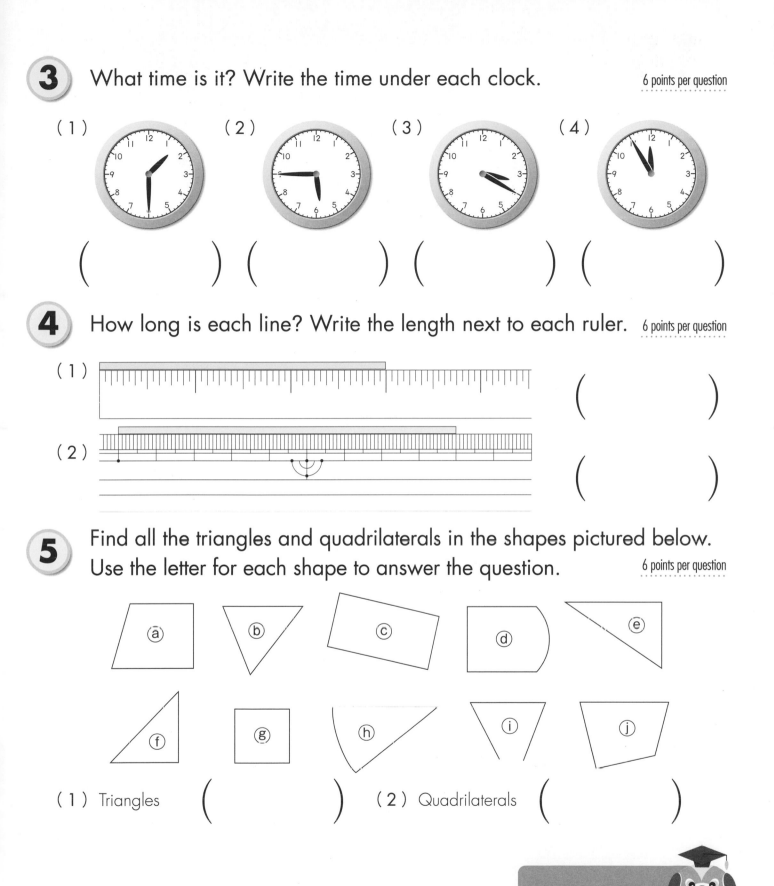 (2) (3) (4)

(   ) (   ) (   ) (   )

4 How long is each line? Write the length next to each ruler. 6 points per question

(1)   (   )

(2)   (   )

5 Find all the triangles and quadrilaterals in the shapes pictured below. Use the letter for each shape to answer the question. 6 points per question

ⓐ   ⓑ   ⓒ   ⓓ   ⓔ

ⓕ   ⓖ   ⓗ   ⓘ   ⓙ

(1) Triangles (   )   (2) Quadrilaterals (   )

You're almost there!

1 Fill in the missing number in each box on the number line below.

4 points per box

(1)

| □ | □ | □ | □ | □ |

9,200 9,300 ↓ 9,500 ↓ 9,700 ↓ 9,900 ↓

(2)

| □ | □ | □ | □ | □ |

7,950 ↓ 7,970 ↓ 7,990 ↓ ↓ 8,020 8,030 ↓

2 Write the appropriate numbers in each box.

4 points per question

(1) The number that is 1 less than **4,000** is [   ] .

(2) The number that is 1 more than **7,009** is [   ] .

(3) The number that is 1 less than **10,000** is [   ] .

3 Write a ✓ under the larger number.

3 points per question

(1) 1,313   1,331
( )   ( )

(2) 4,071   4,091
( )   ( )

(3) 1,001   999
( )   ( )

(4) 8,939   8,937
( )   ( )

4 Draw the long hand on the face of each clock to match the time above.

4 points per question

(1) 10:40　　　　(2) 7:15　　　　(3) 3:55

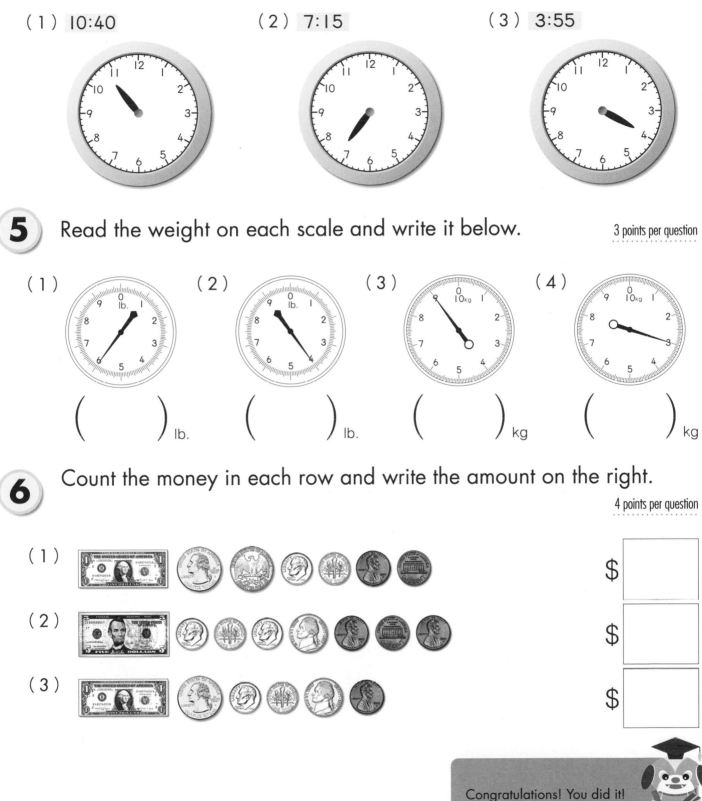

5 Read the weight on each scale and write it below.

3 points per question

(1)　　　　(2)　　　　(3)　　　　(4)

(　　　　) lb.　　(　　　　) lb.　　(　　　　) kg　　(　　　　) kg

6 Count the money in each row and write the amount on the right.

4 points per question

(1)　　$ [　　]

(2)　　$ [　　]

(3)　　$ [　　]

Congratulations! You did it!

1 Review
pp 2,3

1 (1) 86 (2) 48

2 (1) 59 (2) 8, 4 (3) 7

3 (1) 7, 4 (2) 38 (3) 60

4

5 (1) 18 (2) 31
(3) 54 (4) 98

6 (1) ⓐ

2 Review
pp 4,5

1 (1) (From the left) 4, 9, 16, 21, 28
(2) 43, 57, 76, 89, 96

2 (1) 90 (2) 69 (3) 81 (4) 77

3 (1) 32, 23, 13 (2) 96, 86, 68

4 (1) | 36 | 37 | 38 | 39 | 40 | 41 | 42 | 43 |
(2) | 53 | 52 | 51 | 50 | 49 | 48 | 47 | 46 |
(3) | 30 | 40 | 50 | 60 | 70 | 80 | 90 | 100 |
(4) | 22 | 32 | 42 | 52 | 62 | 72 | 82 | 92 |

5 Blue (or blue pencil), 2

6 △, ▢

3 Numbers up to 1,000
pp 6,7

1 (1) 100 (2) 121 (3) 200 (4) 231
(5) 310 (6) 318 (7) 420 (8) 437

2 (1) 536 (2) 863 (3) 273 (4) 445

4 Numbers up to 1,000
pp 8,9

1 (1) 210 (2) 315 (3) 423 (4) 570
(5) 641 (6) 796 (7) 805 (8) 907

2 (1) 2, 5 (2) 4, 8, 5
(3) 8, ones (4) 500
(5) 1,000 (6) 10
(7) 76

3 (1) 2, 4, 3 (2) 3, 0, 7

5 Numbers up to 1,000
pp 10,11

1 (1) (From the left) 10, 60, 110, 190, 240
(2) 320, 380, 430, 460, 520, 580
(3) 100, 400, 600, 700, 900
(4) 870, 910, 940, 960, 990
(5) 761, 776, 784, 799, 817

2 (1) 482 (2) 478 (3) 502 (4) 498

3 (1) 983 (2) 977 (3) 998 (4) 997

6 Numbers up to 1,000
pp 12,13

1 (1) 600 (2) 400 (3) 700 (4) 300
(5) 800 (6) 600 (7) 100 (8) 200
(9) 100 (10) 200

2 (1) 910 (2) 890 (3) 990 (4) 820
(5) 780 (6) 980 (7) 10 (8) 20
(9) 10 (10) 20

7 Numbers up to 1,000
pp 14,15

1 (1) 300 (2) 700
(3) 450 (4) 280
(5) 414 (6) 306
(7) 280 (8) 395
(9) 181 (10) 476
(11) 395 (12) 508

2 (1) 846, 646, 746
(2) 596, 556, 586
(3) 547, 549
(4) 946, 660, 548, 836
(5) 456, 545, 499
(6) 526, 486, 398, 539
(7) 835, 798, 935
(8) 573, 726, 685
(9) 391, 400, 521
(10) 318, 379, 299

8 Numbers up to 10,000 pp 16, 17

1 (1) 1,346 (2) 2,452
(3) 3,278

2 (1) 7,436 (2) 2,463
(3) 2,369 (4) 3,507

9 Numbers up to 10,000 pp 18, 19

1 (1) 2,100 (2) 3,210 (3) 4,315
(4) 6,724 (5) 8,090 (6) 5,109

2 (1) 3, 4 (2) 10,000 (3) 15
(4) 10 (5) thousands, hundreds
(6) hundreds (7) 94

3 (1) 2, 3, 5, 6 (2) 3, 0, 1, 2

10 Numbers up to 10,000 pp 20, 21

1 (1) (From the left) 1,000, 5,000, 7,000, 9,000
(2) 1,100, 2,600, 3,900, 5,400
(3) 4,800, 7,100, 8,700, 9,900
(4) 3,700, 3,900, 4,100, 4,400
(5) 8,600, 8,800, 9,000, 9,300

2 (1) (From the left) 4,910, 4,930, 4,950, 4,970, 4,990
(2) 9,910, 9,940, 9,960, 9,980, 10,000
(3) 8,960, 8,980, 9,000, 9,010, 9,040
(4) 4,991, 4,994, 4,996, 4,998, 5,000
(5) 9,990, 9,992, 9,995, 9,997, 9,999
(6) 6,995, 6,998, 7,000, 7,001, 7,004

11 Numbers up to 10,000 pp 22, 23

1 (1) 1,001 (2) 999 (3) 1,011 (4) 1,009
(5) 1,010 (6) 2,001 (7) 1,999 (8) 2,011
(9) 2,009 (10) 2,010

2 (1) 5,001 (2) 4,999 (3) 5,011 (4) 5,020
(5) 6,999 (6) 9,999 (7) 9,009 (8) 8,000
(9) 5,999 (10) 6,009

3 (1) 10,000 (2) 8,000

12 Numbers up to 10,000 pp 24, 25

1 (1) 3,000 (2) 5,000
(3) 4,500 (4) 9,600
(5) 5,400 (6) 8,530
(7) 2,600 (8) 3,460
(9) 8,090 (10) 6,775
(11) 4,595 (12) 7,209

2 (1) 7,465, 6,465, 8,465
(2) 5,565, 5,965, 5,865
(3) 5,475, 5,485
(4) 5,469, 5,468, 5,467, 5,466
(5) 4,565, 5,290, 5,457, 3,995
(6) 8,350, 7,620, 7,360
(7) 5,730, 7,150, 7,290, 6,850
(8) 6,990, 8,496, 7,014
(9) 6,894, 5,990
(10) 2,975, 3,209, 3,269

13 Telling Time

1 (1) 1:00   (2) 2:00   (3) 3:00
   (4) 5:00   (5) 4:00   (6) 6:00

2 (1) 8:00   (2) 10:00   (3) 11:00
   (4) 12:00

3 (1) 9:30   (2) 11:30
   (3) 1:30   (4) 5:30
   (5) 3:30   (6) 12:30

4 (1) 2:30   (2) 4:30
   (3) 6:30   (4) 10:30

14 Telling Time
pp 28, 29

1 (1) 8:05   (2) 8:10
   (3) 8:15   (4) 8:20
   (5) 8:25   (6) 8:30
   (7) 8:35   (8) 8:40
   (9) 8:45   (10) 8:50
   (11) 8:55   (12) 8:25

2 (1) 8:10   (2) 8:20
   (3) 8:30   (4) 8:40
   (5) 8:50   (6) 8:45
   (7) 8:35   (8) 8:25
   (9) 8:15   (10) 8:05
   (11) 8:55   (12) 8:35
   (13) 8:00

15 Telling Time
pp 30, 31

1 (1) 8:15   (2) 5:15
   (3) 8:20   (4) 11:20
   (5) 8:45   (6) 3:45
   (7) 8:10   (8) 9:10
   (9) 8:55   (10) 10:55
   (11) 8:40   (12) 1:40

2 (1) 6:40   (2) 2:35   (3) 5:50
   (4) 7:25   (5) 4:55   (6) 10:45
   (7) 6:30   (8) 8:05

16 Telling Time
pp 32, 33

1

(1) (2) (3)

(4) (5) (6)

2

(1) (2) (3)

(4)

3

(1) (2) (3)

(4) (5) (6)

4

(1) (2) (3)

(4)

17 Length
pp 34, 35

1 (b)

2 (1) 21   (2) 15   (3) 6

3 (1) 1 in.   (2) 2 in.   (3) 4 in.   (4) 5 in.

18 Length
pp 36,37

1 (1) 2 in., 4 in., 5 in.
(2) 1 in., 2 in., 3 in., 5 in., 6 in.

2 (1) 3 in.   (2) 5 in.

3 (1) 1 in.   (2) 2 in.   (3) 3 in.
(4) 4 in.   (5) 5 in.

19 Length
pp 38,39

1 (1) 1 ft.   (2) 2 ft.   (3) 3 ft.
(4) 4 ft.   (5) 5 ft.
(6) 1 ft. 3 in.   (7) 1 ft. 10 in.
(8) 2 ft. 11 in.   (9) 3 ft. 5 in.
(10) 2 ft. 2 in.

2 (1) 12   (2) 13   (3) 15
(4) 17   (5) 24   (6) 26
(7) 1   (8) 1, 2   (9) 1, 4
(10) 1, 8   (11) 2, 1   (12) 2, 6

20 Length
pp 40,41

1 (1) 1 yd.   (2) 2 yd.   (3) 3 yd.
(4) 4 yd.   (5) 5 yd.   (6) 1 yd. 1 ft.
(7) 1 yd. 2 ft.   (8) 2 yd. 1 ft.
(9) 2 yd. 2 ft.   (10) 5 yd. 1 ft.

2 (1) 3   (2) 4   (3) 6   (4) 8
(5) 10   (6) 14   (7) 1   (8) 1, 2
(9) 2, 1   (10) 3, 1   (11) 4
(12) 5, 1

21 Length
pp 42,43

1 (1) A   (2) D   (3) B   (4) C
(5) A   (6) C   (7) D   (8) A

2 (1) 2 in., 13 in., 22 in.
(2) 3 in., 16 in., 28 in.
(3) 56 in., 69 in., 75 in.
(4) 85 in., 94 in., 107 in.
(5) 190 in., 260 in.
(6) 70 in., 130 in., 210 in.

22 Length
pp 44,45

1 (1) (From the left) 5 cm, 10 cm, 15 cm
(2) 2 cm, 6 cm, 9 cm, 11 cm, 14 cm

2 (1) 7 cm   (2) 13 cm

3 (1) 6 cm   (2) 9 cm   (3) 9 cm
(4) 11 cm   (5) 10 cm

23 Length
pp 46,47

1 (1) 60 cm   (2) 90 cm   (3) 100 cm

2 (1) 2 m   (2) 3 m

3 (1) 1 m 50 cm   (2) 1 m 70 cm
(3) 2 m 80 cm

4 (1) 100 cm   (11) 1 m
(2) 105 cm   (12) 1 m 20 cm
(3) 110 cm   (13) 1 m 50 cm
(4) 130 cm   (14) 1 m 65 cm
(5) 155 cm   (15) 1 m 85 cm
(6) 176 cm   (16) 2 m 10 cm
(7) 195 cm   (17) 2 m 63 cm
(8) 200 cm   (18) 3 m 5 cm
(9) 205 cm   (19) 4 m 15 cm
(10) 247 cm   (20) 5 m 2 cm

24 Weight
pp 48,49

1 (1) 1 lb.   (2) 2 lb.   (3) 3 lb.
(4) 4 lb.   (5) 5 lb.   (6) 6 lb.

2 (1) 2 lb.   (2) 7 lb.   (3) 5 lb.   (4) 9 lb.

3 (1) 1 kg   (2) 2 kg   (3) 3 kg
(4) 4 kg   (5) 5 kg   (6) 6 kg

4 (1) 3 kg   (2) 8 kg   (3) 5 kg   (4) 7 kg

25 Weight
pp 50,51

1 (1) 3 lb.   (2) 8 lb.   (3) 9 lb.

2 (1) 5 kg   (2) 9 kg   (3) 3 kg

3 (1) 1 lb.   (2) 2 kg   (3) 7 kg
    (4) 3 lb.   (5) 9 lb.   (6) 8 kg
    (7) 5 kg   (8) 8 lb.

26 Counting Coins
pp 52,53

1 (1) 80   (2) 90   (3) 95
    (4) 100   (5) 50   (6) 100
    (7) 85   (8) 80   (9) 100
    (10) 90

2 (1) 1   (2) 85   (3) 78
    (4) 1   (5) 90   (6) 1
    (7) 90   (8) 80   (9) 100, 1
    (10) 95

27 Counting Money
pp 54,55

1 (1) 1   (2) 1   (3) 1   (4) 1   (5) 1
    (6) 1   (7) 1   (8) 1   (9) 1   (10) 1

2 (1) 1   (2) 5   (3) 10   (4) 20
    (5) 5   (6) 10   (7) 1   (8) 20

28 Counting Money
pp 56,57

1 (1) 1.01   (2) 1.02   (3) 1.03   (4) 1.04
    (5) 1.05   (6) 1.05   (7) 1.10   (8) 1.25

2 (1) 1.00   (2) 1.01   (3) 1.02   (4) 1.03
    (5) 1.04   (6) 1.05   (7) 1.05   (8) 1.10
    (9) 1.25   (10) 1.50

29 Exchange Unit
pp 58,59

1 (1) 1.00   (2) 0.90   (3) 0.80   (4) 0.70
    (5) 0.60   (6) 0.50   (7) 0.40   (8) 0.30
    (9) 0.20   (10) 0.10

2 (1) 0.20   (2) 0.30   (3) 0.40   (4) 0.60
    (5) 0.70   (6) 0.80   (7) 0.10   (8) 0.90
    (9) 1.00   (10) 0.50

30 Money
pp 60,61

1 (1) ⓐ 1.00 ⓑ 0.80, $1.00 is larger.
    (2) ⓐ 0.90 ⓑ 1.60, $1.60 is larger.
    (3) ⓐ 1.55 ⓑ 1.60, $1.60 is larger.
    (4) ⓐ 1.42 ⓑ 1.46, $1.46 is larger.
    (5) ⓐ 1.55 ⓑ 1.00, $1.55 is larger.

2 (1) ⓐ 1.02 ⓑ 1.75, $1.75 is larger.
    (2) ⓐ 5.50 ⓑ 20.60, $20.60 is larger.
    (3) ⓐ 10.55 ⓑ 1.65, $10.55 is larger.
    (4) ⓐ 20.44 ⓑ 1.41, $20.44 is larger.
    (5) ⓐ 5.55 ⓑ 10.60, $10.60 is larger.

31 Triangles & Quadrilaterals
pp 62,63

1 (1) 2 triangles   (2) 3 triangles   (3) 4 triangles

2 (1) (2) (3) (4) (5) (6) (7)

3 (1) 6 triangles   (2) 5 triangles

4 (1) 4 triangles   (2) 10 triangles
    (3) 9 triangles

(32) Triangles & Quadrilaterals   pp 64,65

1 (1) 3 sticks   (2) 4 sticks   (3) 7 sticks
(4) 6 sticks   (5) 5 sticks   (6) 7 sticks

2 (1) 12 sticks   (2) 16 sticks

3 (1) ⓐ, ⓒ, ⓓ, ⓖ
(2) ⓑ, ⓔ, ⓕ

4 (1) ⓐ, ⓖ   (2) ⓑ, ⓔ, ⓕ, ⓘ

(33) Triangles & Quadrilaterals   pp 66,67

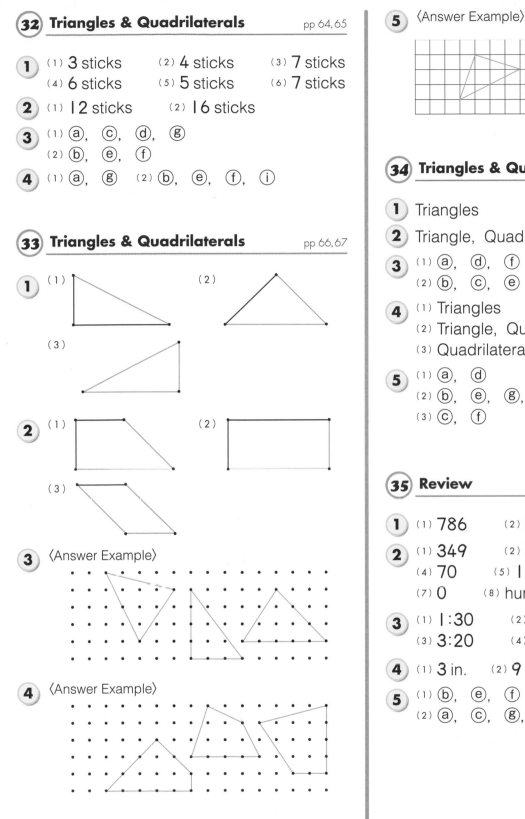

1 (1)   (2)
(3)

2 (1)   (2)
(3)

3 ⟨Answer Example⟩

4 ⟨Answer Example⟩

(5) ⟨Answer Example⟩

(34) Triangles & Quadrilaterals   pp 68,69

1 Triangles

2 Triangle, Quadrilateral

3 (1) ⓐ, ⓓ, ⓕ
(2) ⓑ, ⓒ, ⓔ

4 (1) Triangles
(2) Triangle, Quadrilateral
(3) Quadrilaterals

5 (1) ⓐ, ⓓ
(2) ⓑ, ⓔ, ⓖ, ⓗ
(3) ⓒ, ⓕ

(35) Review   pp 70,71

1 (1) 786   (2) 5,462

2 (1) 349   (2) 5,020   (3) 1,000
(4) 70   (5) 10,000   (6) 49
(7) 0   (8) hundreds

3 (1) 1:30   (2) 5:45
(3) 3:20   (4) 11:55

4 (1) 3 in.   (2) 9 cm

5 (1) ⓑ, ⓔ, ⓕ
(2) ⓐ, ⓒ, ⓖ, ⓙ

1 (1) (From the left) 9,100, 9,400, 9,600, 9,800, 10,000

(2) 7,960, 7,980, 8,000, 8,010, 8,040

2 (1) 3,999 (2) 7,010 (3) 9,999

3 (1) 1,331
(2) 4,091
(3) 1,001
(4) 8,939

4 (1) (2)  (3)

5 (1) 6 lb. (2) 4 lb. (3) 9 kg (4) 3 kg

6 (1) 1.72 (2) 5.38
(3) 1.51